Mineral Processing

Technologies, Challenges and Perspectives

About the Centre

The Centre for Science and Technology of the Non-Aligned and Other Developing Countries (NAM S&T Centre) is an inter-governmental organisation with a membership of 48 countries spread over Asia, Africa, Middle East and Latin America. Besides this, 12 S&T agencies and academic/research institutions of Bolivia, Botswana, Brazil, India, Nigeria and Turkey are the members of the S&T-Industry Network of the Centre. The Centre was set up in 1989 to promote South-South cooperation through mutually beneficial partnerships among scientists and technologists and scientific organisations in developing countries. It implements a variety of programmes including international workshops, meetings, roundtables, training courses and collaborative projects and brings out scientific publications, including a quarterly Newsletter. It is also implementing 7 Fellowship schemes, namely, NAM S&T Centre Research Fellowship, Joint NAM S&T Centre – ICCBS Karachi Fellowship, Joint CSIR/CFTRI (Diamond Jubilee) - NAM S&T Centre Fellowship, Joint NAM S&T Centre – ZMT Bremen Fellowship, Research Training Fellowship for Developing Country Scientists (RTF-DCS), NAM S&T Centre – U2ACN2 Research Associateship in Nanosciences and Nanotechnology and Joint NAM S&T Centre – DST (South Africa) Training Fellowship on Minerals Processing and Beneficiation in Indian institutions. These activities provide, among others, the opportunity for scientist-to-scientist contact and interaction, training and expert assistance, familiarising the scientific community on the latest developments and techniques in the subject areas, and identification of technologies for transfer between member countries. The Centre has so far brought out 71 publications and has organised 100 international workshops and training programmes.

For further details, please visit www.namstct.org or write to the Director General, NAM S&T Centre, Core 6A, 2nd Floor, India Habitat Centre, Lodhi Road, New Delhi-110003, India (Phone: +91-11-24645134/24644974; Fax: +91-11-24644973; E-mail: namstcentre@gmail.com; namstct@bol.net.in).

Mineral Processing

Technologies, Challenges and Perspectives

— *Editor* —

G. Padmanabham

CENTRE FOR SCIENCE & TECHNOLOGY OF THE
NON-ALIGNED AND OTHER DEVELOPING COUNTRIES
(NAM S&T CENTRE)

2017
DAYA PUBLISHING HOUSE®
A Division of
ASTRAL INTERNATIONAL PVT. LTD.
New Delhi – 110 002

Publisher's Note:

Cataloging in Publication Data--DK
Courtesy: D.K. Agencies (P) Ltd. <docinfo@dkagencies.com>

International Workshop on 'Mineral Processing and Beneficiation' (3rd : 2014 : Harare, Zimbabwe)
Mineral processing : technologies, challenges and perspectives / editor, G. Padmanabham.
pages cm
"Centre for Science & Technology of the Non-Aligned and Other Developing Countries (Nam S&T Centre)."
ISBN 9789387057227 (International Edition)

1. Ore-dressing--Developing countries--Congresses. 2. Ore-dressing--Economic aspects--Developing countries--Congresses. 3. Mineral industries--Developing countries--Congresses. 4. Mineral industries--Technological innovations--Developing countries--Congresses. 5. Mineral industries--Government policy--Developing countries--Congresses. I. Padmanabham, G., editor. II. Zimbabwe. Ministry of Higher and Tertiary Education, Science and Technology Development, organizer. III. Zimbabwe. Ministry of Mines and Mining, organizer. IV. Centre for Science and Technology of the Non-Aligned and Other Developing Countries. V. Title.

TN500.M56 2014   DDC 622.7   23

Centre for Science and Technology of the Non-Aligned and Other Developing Countries (NAM S&T Centre)
Core-6A, 2nd Floor, India Habitat Centre, Lodhi Road,
New Delhi-110 003 (India)
Phone: +91-11-24644974, 24645134, Fax: +91-11-24644973
E-mail: namstct@gmail.com
Website: www.namstct.org

Published by   : **Daya Publishing House®**
                          A Division of
                          Astral International Pvt. Ltd.
                          –ISO 9001:2015 Certified Company –
                          4736/23, Ansari Road, Darya Ganj
                          New Delhi-110 002
                          Ph. 011-43549197, 23278134
                          E-mail: info@astralint.com
                          Website: www.astralint.com

Foreword

It is well recognised by the economists around the globe that along with physical and human capital, natural resources arealso an important asset with a significant role in the economic development of any nation or region.The natural resources of a country can appreciably contribute to its economic development through increasing production and manufacturing, securing energy supply, increasing export revenues and reducing costs for local businesses and households. Natural resources include, among other, the land, forest, minerals, climate, water *etc.*

Minerals refer to the inorganic substances which are produced in nature and are obtained in some combined state.Mining is the process of extracting out mineral from the earth *e.g.*, quartz, calcite, feldspar *etc.*Beneficiation, or value-added processing, involves the transformation of a primary material (produced by mining and extraction processes) to a more finished product, which has a higher sales value. In most of the developing countries these natural resources either remain unutilised or underutilised due to mainly the lack of relevant advanced skills, infrastructure and modern technologies. There is therefore a need for the policy makers, researchers and other stake holders involved in exploitation of mineral processing and beneficiation to play a role in arresting this situation.

I am thrilled by the initiatives of the Centre for Science and Technology of the Non-Aligned and Other Developing Countries (NAM S&T Centre) that recognises the importance of Mineral Processing and Beneficiation and had taken a first positive step by organisinga series of three international programmes on this issue. The first programme was the International workshop on 'Mineral Resources and Development' in July 2004 at Kerman, Iran in association with Shahid Bahonar University of Kerman. The subsequent one was the International Workshop on 'Minerals Processing and Beneficiation' organised in September 2012 in Johannesburg, South Africa jointly with the Department of Science and Technology of South Africa. The 3rd International Workshop on 'Mineral Processing and Beneficiation'took place in Harare, Zimbabwe during 11-14 September 2014 jointly with the Ministry of Higher and Tertiary Education, Science and Technology

Development and the Ministry of Mines and Mining Development of the Republic of Zimbabwe.

I commend the NAM S&T Centre for now bringing out this highly valuable third publication in the series on the topic of '**Minerals Processing and Beneficiation**'. The publication comprisesof 14 papers from 12 developing countries. I am confident that the book will be an asset to the policy makers and researchers, NGOs and all others who are associated with mineral processing and beneficiation activities especially in member states of the NAM S&T Centre and those in other developing countries.

Abiel Mngomezulu,
CEO and President, MINTEK,
Private Bag X3015, Randburg,
South Africa

Preface

Minerals processing and beneficiation technologies are key to effective exploitation of mineral resources available, without adverse ecological effects. The local conditions, nature and resource size determine which one is an appropriate technology for a given need. At the same time, building an overall ecosystem, which makes use of the technology, needs to be addressed through suitable policies. In this context, each country has different type of challenges to be addressed and exchange of experiences and knowledge between experts from different nations will go a long way in effective utilisation of mineral resources through appropriate technologies.

This book is a compilation of such information and perspectives shared by speakers from 12 countries on mineral resources, problems in processing and beneficiation, localised solutions adopted and socioeconomic challenges that different countries are faced with or making efforts to address.

With respect to Iron and Steel sector, Dr. Ratnakar Singh of National Metallurgical Laboratory of India presented the innovative processes developed and adopted for reducing alumina and silica impurities. For example, heavy media separation for decreasing the laterite material in the lumps, gravity-cum-magnetic separation process for processing of iron ore fines, hydrocycloning and gravity separation *etc.*, were found to be useful. The article of Matinde and Mokoni presented a detailed gap analysis with respect to important factors affecting the technological and innovation capabilities in Zimbabwe vis-à-vis the Iron and Steel sector and possible actions. The Zimbabwean speaker, Mpofu highlighted the importance of policing the mineral resources.

It is interesting to note the efforts being made by countries like Mauritius, which have very limited mineral resources. Their recent investigation on their local soils showed the presence of Kaolinite and Montmorillonite. On the contrary, mineral rich countries like Myanamar with lead-zinc-copper, tin-tungsten, gold-silver, and gemstones are looking for new and efficient technologies to commercially exploit them in joint venture mode through its open door policy. Similarly, Nigeria is well endowed with mineral resources, which when utilised well can be a major economic driver. Adewole *et al.,* from RMRDC of Nigeria reported the large number of minerals that are available and also highlighted the need to evolve strategies including technological interventions for value addition and consequent economic benefits. The huge investment opportunities in the solid mineral sector in Nigeria were highlighted.

On the precious metals side, Yudi *et al.,* have brought out the harmful effects of artisanal practices in gold mining using mercury and highlighted the Indonesian action plan for mercury elimination through development and dissemination of mercury-free technology by establishing an Indonesian Centre for Artisanal Mining (INCAM). The Platinum Group Metals (PGMs) iridium, osmium, palladium, platinum, rhodium, and ruthenium come from South Africa, Russia, North America and Zimbabwe. PGMs have wide applications considering their catalytic properties, wear and tarnish resistance characteristics, resistance to chemical attack, excellent high-temperature characteristics and stable electrical properties. Tony Nyakudarika brought out the current conditions and thoughts in terms of sustainable PGM mining in Zimbabwe.

Sri Lanka is endowed with graphite, high purity quartz, mineral sand, limestone, dolomite, clay minerals, feldspar, apatite (rock phosphate) gemstones and mica. Minerals such as graphite, mineral sand, vein quartz and mica are exported after some beneficiation and preliminary processing. Sri Lanka's intentions of value addition by introducing advanced mineral processing and beneficiation methods to increase income from exports are highlighted by D. Sajjana de Silva from Sri Lankan Geological Survey and Mines Bureau. For example, froth flotation for graphite mining of low grade and complex ore bodies, use of hydrocyclones for separation of coarse clay particles and fine sand during Kaolin refining *etc.,* were presented.

On the policy side, commodity exchange places as an effective mechanism for better monetization of minerals and governmental interventions required for efficient use of the resources was discussed by Uganda. The cost-benefit analysis of bauxite exploitation projects in the Tay Nguyen area in Vietnam considering various socio-economic aspects is an interesting study giving insights into effects of mining and possible benefits.

Apart from the above, some very interesting scientific research results were presented. Research on use of natural Malaysian silica sand as the SiO_2 raw material for producing (KAl_2SiO_6) glass-ceramics is reported by Malek *et al.,* The leucite glass-ceramics so developed were found to be translucent with a flexural strength values (80–175MPa) comparable to commercial products. The bio-inertness and non-cytotoxicity of this ceramic can possibly find restorative dental applications. Carrollite, known to be the main source of Cobalt, from the Zambian copper-cobalt bearing ores is a strategic mineral. Detailed experimental work to induce surface hydrophobicity and subsequent effects yielding recovery of 95.21 per cent was reported.

In summary, this book covers interesting perspectives on the mining and minerals sector activities in 12 countries including the technical as well as policy aspects and suggestions to maximise the economic value. I am sure the book provides very good insights for all stakeholders in the field of minerals processing and beneficiation, not only in the countries who participated in this NAM S&T Workshop but also other countries with interests in this field.

Dr. G. Padmanabham

Director

International Advanced

Research Centre for

Powder Metallurgy &

New Materials (ARCI)

Hyderabad-500 005

India

Introduction

Minerals resources form an important component of the nation's economic prosperity. Most of the emerging economy countries are minerals rich, yet find it extremely difficult to do the value addition by converting the precious raw material to industrial product. Non-accessibility to appropriate knowhow, inability to secure necessary financial resources and non-availability of proficient and skilled manpower are some of the impediments that force them to sell their mineral deposits in the raw form without adequate financial dividend. The developing countries need to focus their efforts and build up strategies for exploitation of their resources through modern and innovative researches and technological advances within an environment of sustainability consciousness in new world trade regime. In this regard, the policy makers, scientists, technologists, academics, industry professionals and other stakeholders should come together and assess ways and means to realise the potential of the mineral resource base and its contribution to industrial development of their countries.

In the above context, the Centre for Science and Technology of the Non-Aligned and Other Developing Countries (NAM S&T Centre) had organised in the past a series of international workshops on 'Minerals Processing and Beneficiation' respectively, (i) at Kerman, Iran in July 2004 in association with Shahid Bahonar University of Kerman; (ii) at Johannesburg, South Africa in September 2012 jointly with the Department of Science and Technology (DST) of the Republic of South Africa; and (iii) at Harare, Zimbabwe during 11-14 September 2014 jointly with the Ministry of Higher and Tertiary Education, Science and Technology Development and the Ministry of Mines and Mining Development of the Republic of Zimbabwe. The latest of these scientific events was inaugurated by H.E. Dr. Robert G. Mugabe, Honourable President and Head of the State and the Government of the Republic of Zimbabwe and was attended by 110 participants from 15 countries representing Afghanistan, Guyana, Indonesia, Iran, Malaysia, Mauritius, Myanmar, Nigeria,

South Africa, Sri Lanka, Tanzania, Uganda, Vietnam and Zambia, as well as the host country Zimbabwe, from which there were 89 delegates.

As the immediate follow up, a 'Harare Declaration – 2014 on Mineral Processing and Beneficiation' with a set of recommendations was unanimously adopted by the delegates at the conclusion of the workshop. Yet in order not to lose the acquired knowledge and sharing of experience from this workshop, we have brought out the present book, which comprises 14 scientific and technical papers contributed by the experts from 12 countries.

I would like to express my gratitude to Dr. G. Padmanabham, Director, International Advanced Research Centre for Powder Metallurgy and New Materials (ARCI), Hyderabad, India, who is a very well renowned expert on Materials Joining, Laser Processing of Materials and Technology Transfer and Commercialisation, for technical editing of the present volume. I also appreciate the efforts put in by the entire team of the NAM S&T Centre, especially by Dr. (Mrs.) Kavita Mehra, Mr. M. Bandopadhyay, Ms. Geeta and Mr. Pankaj Buttan in compiling and checking the manuscripts, liaising with the authors, cover page designing, proof reading, formatting and taking all the necessary actions in giving a shape to this book.

I trust that this book will be useful to all the stakeholders for gaining insight on various aspects of Mineral Processing and Beneficiation, from researchers to policy makers and government officials to the industry personnel in the developing countries.

Prof. Dr. Arun P. Kulshreshtha
Director General,
NAM S&T Centre

Contents

Chapter 1

Beneficiation of Iron Ores for Iron and Steel Making: Problem and Prospects

Ratnakar Singh

Chief Scientist and Head, Mineral Processing Division,
CSIR-National Metallurgical Laboratory,
(Council of Scientific and Industrial Research),
Jamshedpur – 831 007, India,
E-mail: rs@nmlindia.org

ABSTRACT

India is endowed with large reserves of iron ores which is the basic raw material for iron and steel making. Although Indian iron ore is rich in iron but it contains high amount of alumina and phosphorus which are not favourable for efficient operation of blast furnace. The present paper deals with the available resources and characteristics of Indian iron ores, the deleterious effects of high alumina, industrial beneficiation practices and their limitations. In the light of recent developments and rich experience of CSIR-NML on beneficiation of iron ores, the process options for lowering alumina are discussed and salient results on new processing techniques are presented. From the point of view of increasing demand; resource conservation and environment protection; total beneficiation of iron ores covering lumps; fines and slimes and the utilisation of low grade ores; blue dust; dumped fines and slimes through beneficiation and agglomeration are emphasised.

Keywords: Iron ore, Characteristics, High alumina, Beneficiation, Sintering, Pelletization.

INTRODUCTION

The world reserves of iron ore are estimated to be around 170 billion tonnes (Anonymous, 2012). The principal minerals of iron are the oxides (hematite and magnetite), hydroxide (limonite and goethite) and carbonate (siderite). In nature the commercial deposits are mostly of bedded type, although deposits of magmatic, contact metasomatic and of a replacement nature also exist. In many cases, ground water circulation and weathering have resulted in concentration of the ore from primary sources. The major iron ore producing countries in the world are the Australia, Brazil, Canada, China, India, USA, Russia, Kazakhstan, South Africa, Ukraine and Sweden. Pre Cambrian banded iron formations containing 30 per cent or more of iron are the predominant sources of iron.

The primary product from iron ore is pig iron, produced by smelting in a blast furnace. The requirements of economic blast furnace operation set the specifications for marketable iron ores and agglomerates. The requirements differ in different parts of the world and also to some extent from furnace to furnace depending upon the quality of fluxes and fuels available and on the end use of pig iron. The principal impurities associated with iron ores are silica, alumina, sulphur and phosphorous. In general, as per the price-quality consideration, marketable ores should contain maximum iron with minimum gangue. In particular, one looks for minimum contents of SiO_2 and Al_2O_3 because of the ratio of acid (SiO_2, Al_2O_3) to basic (CaO, MgO) constituents of the gangue has influence on furnace operation. It is desirable to burden a furnace to attain minimum slag volume, consistent with effective fluxing of gangue and with optimum elimination of elements such as sulphur and alkalies in the slag (Weiss, 1985). It may be worth mentioning that for hematite ores, typical lump grades contain over 62.5 per cent of iron Pellets for blast furnace charge contain about 65 per cent Fe and <5 per cent SiO_2 while for direct reduction, average iron is more than 67 per cent with silica plus alumina less than 2 per cent. Due to the demand of ores with higher contents of iron and low level of impurities, today in majority of the cases the run-of-mine ore is processed to varying degrees before the shipment.

India is one of leading producers of iron ores in the world and it can meet the growing demand of iron and steel industry in the country and also sustain considerable foreign trade. As per the National Steel Policy, India will be producing 200 million tonnes of steel by 2019-20. This would require about 350 million tonnes of iron ore for domestic steel production in addition to meeting the demand on foreign trade. Indian iron ores are rich in iron content but generally contain high alumina which is undesirable for its use for iron making. The high alumina of iron ores is not favourable for the efficient operation of blast furnace. For an efficient operation of blast furnace, the alumina/silica ratio in the ore feed should be <1.0 and alumina should be <2 per cent. But most of the Indian hematitic ores, which are used for iron and steel making, show a reverse trend. Alumina bearing minerals are more or less intergrown with the iron oxide minerals and in general the liberation takes place in the fine size ranges. Most of the iron ore mines in India are operated by selective mining for maintaining a high grade (60 per cent Fe). With the availability of high grade ores and for the economic reasons, a simple washing scheme has been industrially adopted for beneficiation of Indian iron ores. It helps in removal of adhering clay and silica to produce free flowing lumps and sand. But it has limitations in reduction of impurities and sometimes desired alumina/

silica ratio is not achieved. The present beneficiation practice also generates fines and slimes varying from 10 to 25 per cent. These slimes are accumulated along iron ore mines causing environmental pollution related problems besides losses of iron values. Considering the increasing demand of quality iron ores, limitations of the current beneficiation scheme in particular for processing of low grade ores and the conservation of iron ore resources, there is need to develop and adopt suitable technology for beneficiation of Indian iron ores.

CSIR-National Metallurgical Laboratory (CSIR-NML) has been a pioneer in mineral processing and involved in beneficiation of iron ores from entire geographical region of the country and as per the industrial needs it has provided know-how for setting up iron ore beneficiation and agglomeration plants (Singh *et al.*, 2004). CSIR-NML has also carried out detailed studies and developed processes for beneficiation and agglomeration of low grade iron ore samples from abroad such as Egypt, Syria, Nepal, Mali, Morocco, Ethiopia, Kazakhastan, Ukraine, *etc*. In this paper an attempt has been made to discuss the problem of concentration of Indian iron ores with high alumina content. Based on the extensive studies undertaken at CSIR-NML the results on the development of processes for beneficiation of iron ores are presented.

Indian Iron Ore: Resources and its Characteristics

With *in-situ* reserves of over 28 billion tonnes, India is one of the leading producers and exporter of iron ore in the world. Hematite and magnetite are the two important iron ores, of which hematite is the major resource. Hematite type Indian deposits belong to the precambrian iron ore series and the ore being mainly confined in banded iron ore formations. Magnetite type ore occurs either as igneous origin or the metamorphosed banded magnetite-silica formation possibly of sedimentary origin. About 60 per cent of hematite ore deposits are found in Eastern Sector while about 80 per cent magnetite ore deposits occur in Southern Sector, specially in Karnataka. The geographical distribution of banded iron formation in India is shown in Table 1.1.

Table 1.1: Distribution of Iron Ore Resources in India

Zone A	Singhbhum, Keonjhar, Cuttack (Jharkhand and Orissa)
Zone B	Bastar, Durg, Chandrapur (Madhya Pradesh and Maharashtra)
Zone C	Bellar-Hospet belt (Karnataka)
Zone D	Goa, Ratnagiri, North Kanara
Zone E	Metamorphosed banded iron formation along the West Coast (Karnataka, Kerala *etc.*)

The hematite ores range from massive type through a porous laminated type to a fine soft powder and are generally classified under the following categories : (a) Massive, (b) Laminated - soft and hard, (c) Lateritic and (d) Powdery/Blue Dust. The chemical analysis of the different varieties of iron ore is shown in Table 1.2 while Table 1.3 summarises the mineralogical characteristics (Singh *et al.*, 2004). As it can be seen from Table 1.3, lateritic ores show high alumina content while blue dust is very rich with iron content and low alumina and silica impurities but it is found in powder form.

Table 1.2: Chemical Analysis of Hematitic Ore

Ore Type	Fe Per cent	SiO_2 Per cent	Al_2O_3 Per cent
Massive	63.2-69.0	0.34-4.19	0.50-3.66
Hard Laminated	56.8-66.6	0.81-5.73	1.00-7.14
Soft Laminated	57.0-65.5	1.20-8.60	1.10-9.00
Lateritic	56.0-61.5	1.00-6.88	3.72-10.85
Blue Dust	64.0-69.0	0.64-2.12	0.35-2.49

As found from the petrological studies and shown in Table 1.3, hematitic ore basically consists of varying amounts of hematite, goethite, martite and magnetite in association with quartz and clay as the gangue forming minerals. Although Indian iron ores are considered to be rich in iron content but they contain high alumina. Due to preferential association of iron oxide mineral with finely disseminated alumina bearing minerals (kaolinite, gibbsite) than with siliceous minerals, the Al_2O_3/SiO_2 ratio is generally more than 1. Alumina is mainly contributed by clay (kaolinite, montmorillonite, illite, alunite), lateritic material and gibbsite and some alumina occurs as solid solution in iron oxide minerals *viz.*, goethite and limonite (Pradip, 1995; Singh, 2004). Silica is mainly contributed by quartz and the associated clay. The adverse ratio of Al_2O_3/SiO_2 in Indian iron ores has always been a problem in improving the blast furnace productivity.

Table 1.3: Mineralogical Characteristics of Hematitic Ores

Ore Type	Iron Bearing Minerals	Gangue Minerals	Other Features
Massive	Hematite, Goethite, Martite and Magnetite	Quartz, Clay	Steel grey in colour, Specific gravity >5, High crushing strength
Laminated	Hematite, Goethite, Limonite	Clay, Gibbsite, Quartz, Chert	Laminated structure, Specific gravity 4.2-4.7
Lateritic	Goethite, Limonite, Hematite, Ochre	Clay, Gibbsite Silica, Ochre	Dull lusture, Rich in alumina, Friable nature
Blue Dust	Hematite,Goethite	Quartz, Clay	Generally blue/dark black or cherry red in colour, Powder form, Low alumina

Studies show that even below 75 micron, significant portion of alumina minerals remain locked with iron oxide minerals. Results on liberation studies carried out on two low grade iron ore samples, namely S1 and S2, crushed to 2 mm are shown in Figure 1.1. As it can be seen, liberation of iron bearing minerals increases with decrease in particle size and about 80 per cent liberation is achieved below 100 micron. Magnetite ores are of low grade (30-40 per cent Fe) with quartz as the main gangue. These ores are finely pulverized and need even finer grinding for release of valued minerals from gangues.

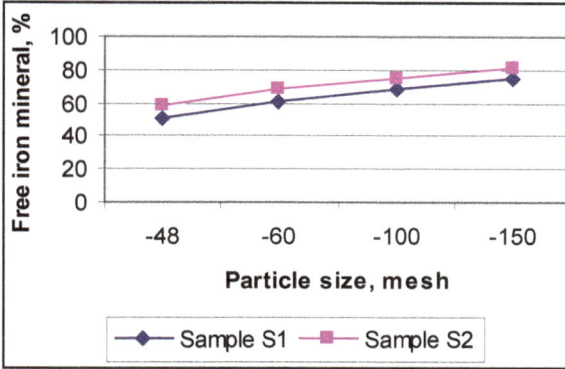

Figure 1.1. Liberation of Iron Bearing Minerals with Decreasing Particle Size

Quality of Iron Ores and Deleterious Effects of Alumina

In India, blast furnace route dominates the bulk of steel production and the performance of blast furnace is greatly influenced by the quality of raw materials. The broad specifications of iron ore in for SAIL plants have been as follows : 62-65 per cent, Al_2O_3 2.5-3.5 per cent, Al_2O_3 + SiO_2 5-6 per cent, Al_2O_3/SiO_2 1.0-1.2 per cent (Sahu, 1995). However, for achieving better blast furnace productivity and slag flowability, the iron ore should contain iron as high as possible and Al_2O_3/SiO_2 ratio <1.0 and Al_2O_3 <2.0 per cent. Normally alumina content in iron ores lies in the range of 2-3 per cent in the lump and 4-6 per cent in the fines used for sinter making as against world practice of alumina below 1 per cent. Sinter produced from such iron ore fines (-10 mm) is extremely high in alumina compared to sinter from other countries where the alumina in the burden seldom exceeds 2 per cent. Alumina has an adverse effect on sinter properties like Reduction Degradation Index (RDI). Burden in the Indian blast furnaces is slow moving with a longer residence time for sinter/ore. Under these circumstances RDI is even more important. It is estimated that the drop in sinter alumina from 3.1 to 2.5 per cent can improve RDI of sinter by at least 6 points. This would in turn lower blast furnace coke rate by about 14 kg/thm and increase productivity by about 3 per cent.

The lumps containing about 2 per cent and fines containing about 4 per cent alumina, coupled with alumina input through coke and other raw materials result in 22-26 per cent alumina content in the blast furnace slag. Blast furnace slag at this level of alumina is generally viscous. As a consequence, problems associated with blast furnace permeability are frequently encountered. This impairs the upward flow of reducing gases and decreases reduction kinetics of iron oxide. Viscous slag is also not conducive to efficient desulphurising. High alumina loading in Indian blast furnaces has prompted the operators to resort to 9-10 per cent MgO in blast furnace slag to make it fluid (Rao *et al.*, 1995). Thus lowering of alumina content in iron ore is essential for an efficient operation of blast furnace.

Present Scenario of Iron Ore Benefication in India

Since its inception in early fifties, CSIR-NML has been engaged in R&D in beneficiation and agglomeration of iron ores and fines for utilisation in iron and steel making. Detailed characterisation, laboratory and pilot scale beneficiation and agglomeration studies were undertaken on iron ore samples obtained from various captive mines supplying iron ores to the various public as well as private sector undertakings such as SAIL, TISCO, NMDC *etc.* Based upon these studies, as per the industrial need know-how was provided for installation of various iron ore washing and agglomeration plants in the country (Table 1.4).

Table 1.4: Iron Ore Washing and Agglomeration Plants Based on CSIR-NML's know-how

Iron Ore Mines	Steel Plant/Purpose
Noamundi	TISCO
Dalli-Rajhara	Bhilai Steel Plant
Barsua	Rourkela Steel Plant
Gua	Indian Iron and Steel Co.
Bolani	Durgapur Steel Plant
Kiriburu	Bokaro Steel Plant
Bailadila	Export
Donimalai	Export
Kudremukh	Export
Kanjamalai	Salem Steel Plant
Daitari-Ganndhamardan	Paradip Steel Plant
Goa Iron ore	Export

In view of availability of good grade ores and liberal product quality and the economic reasons, most of the Indian iron ore mines are operated by selective mining for maintaining a high grade of iron (Fe>60 per cent). There are 13 ore processing plants in the country (excluding Goa and Kudremukh) and the ore is either treated in dry or wet circuits. The dry circuit consists of crushing and screening, while the wet circuit consists of crushing, scrubbing, wet screening and classification. The prime function of beneficiation is to improve the iron content and to decrease the alumina/iron ratio.

Scrubbing and/or wet screening are essential for aluminous and sticky ores. Washing ensures separation of clayey matter from lumps and fines and improves the handling properties, particularly in monsoon season. The sticky nature of the ore due to the presence of clayey matter causes serious problem in transportation. The final products are lumps (generally 10-40 mm in size), classifier sand as sinter feed (0.15-10 mm) and -0.15 mm slime. Hydrocyclones are used to recover iron values from slime. With rich ROM feed to the washing circuit, hydrocyclone underflow can be processed through filters or slow speed classifiers to recover high grade microfines. Slow speed classifiers in combination with hydrocyclones can either be utilised for dewatering or recovering microfines.

The treatment lump size varies from -30 + 10 mm to -150 + 10 mm from mines to mines. For mixed hematite and magnetite ores, the beneficiation essentially involves crushing, grinding, magnetic separation, gravity separation by spirals and final concentration by flotation.

Limitations of Conventional Beneficiation Practice and Scope for Improvement

As mentioned above, the conventional scheme of iron ore beneficiation and the reduction of alumina broadly comprises crushing the ore to the required size followed by scrubbing and/or wet screening and classification to separate slimes from fines. The scheme was developed for processing of high grade ores with comparatively liberal product's quality norms and targeted for utilisation of lumps and fines only. But with the changed scenario, particularly considering the increasing demand for high grade iron ore for iron making, lowering of alumina content to the specified level and the need of processing of low grade feed, the conventional beneficiation circuits have limitations. By this process, although the adhering clay matter is removed, the Al_2O_3 content is not significantly lowered. One of the most important aspects in the beneficiation of Indian iron ores is their complex nature from the standpoint of elimination of alumina. The general mode of occurrence of aluminous minerals in the ore is as coating on lumps, as cavity fillings or as lateritic material. Laterites containing over 8 per cent alumina exist as a constituent phase along with iron oxide. Quartz occurs as bands with hematite in BHQ/BHJ as inclusions, adherent and vug fillings. By washing, alumina present as clayey material as well as fine silica can be partially removed, whereas laterit is not affected by washing. Thus there is preferential removal of silica over alumina, which naturally increases the alumina/silica ratio in the washed product. The present iron ore washing scheme also generates slimes (10-25 per cent by weight) with relatively high alumina content. These slimes stored in tailings ponds pose environmental pollution problems.

Considering the need of controlling the alumina content of the charge to blast furnace coupled with the complexity of the beneficiation of Indian iron ores, considerable R&D efforts have been made at CSIR-NML to develop suitable technology for meeting the requirement of low alumina content of iron ore. In order to meet the requirement of low alumina content of iron ore and also taking care of environment pollution problems and the need of resource conservation, total processing of iron ore covering lumps, fines and slimes is recommended. In view of the gradual shift from lumps to fines in Indian blast furnace and the recent developments in the area of beneficiation, there is enough scope to design an integrated flow-sheet consisting of processing of lumps, fines and slimes. In particular, gravity based beneficiation techniques like heavy media separation, jigging, spiralling, enhanced gravity separation, magnetic separation, flotation and column flotation have enough potential for treating Indian iron ore lumps, fines and slimes. Blue dust having high iron and low gangue contents and beneficiated fines need to be utilised by sintering. Studies have been carried out at CSIR-NML and other laboratories in the country on these lines. Depending upon the mineralogical, liberation and beneficiation characteristics, the process can be tailored to suit the specific iron ore deposit leading to the optimum results with respect to yield and the quality of the concentrate.

R&D and Salient Results on Benefication of Indian Iron Ores

As mentioned in the previous sections, research and development studies were undertaken at CSIR-NML on beneficiation of iron ore samples to address the problem of Indian Iron and Steel Industry. Recently CSIR-NML has carried out comprehensive studies on different varieties of iron ore samples obtained from Barsua Mines, Daitari Gandamardhan mines, iron ore fines from Hospet and Barbil, Chiria and other mines in the country as well as many samples from abroad to develop process flow-sheet for reducing alumina and silica impurities. Studies have also been undertaken to develop process for preparation of iron ore super-concentrate for high-tech applications and pelletization and sintering of iron ore fines. Further studies are under progress on developing technologies for processing low grade and waste dump iron ore to a high grade concentrate to be used as feed to pellet for DRI plant. Automated flotation column developed by CSIR-NML has been used for studying beneficiation of low grade iron ores and fines (Prabhakar *et al.*, 1997). These studies have led to the development of innovative technologies for beneficiation and agglomeration of low and lean grade iron ores (Dey *et al.*, 2012; Singh *et al.*, 2014). Salient results are discussed below :

Beneficiation of Iron Ore Lumps

So far as the beneficiation of iron ore lumps is concerned, due to limitation on size and liberation, we have limited choice. Normally lumps can be beneficiated by scrubbing and washing and heavy media separation. Jigging is also reported to be applicable to the processing of iron ore lumps. The air pulsated jig has given good response for the processing of lumps. Results on processing of typical iron ore samples are described below :

Washing :

Studies were carried out on reduction of alumina in iron ores from various sources by scrubbing and washing technique. Typical results on washing of iron ores from selected sources are shown in Table 1.5. As it is evident from the data, washing helps in lowering of alumina and silica from the iron ore lumps. This is mainly due to removal of adhering clay or cavity filling and fine silica.

Heavy Media Separation

While washing removes the adhering clay or cavity filling and fine silica, the laterites remain unchanged and there is preferential removal of silica over alumina which naturally increases the alumina/silica ratio in the washed product. It has been observed that gravity methods like heavy media separation can be successfully employed to decrease this laterite material of the ores due to the difference in specific gravity between the iron ore minerals and the laterite. There is a preferential rejection of the alumina over silica when lateritic materials are rejected.

Table 1.5: Typical Results on Washing of Indian Iron Ore Samples

Source	Feed Size, mm	Feed Assay, Per cent			Product Assay, Per cent		
		Fe	SiO$_2$	Al$_2$O$_3$	Fe	SiO$_2$	Al$_2$O$_3$
Joda	−50	61.6	2.70	4.30	62.80	1.84	3.73
	−25				63.30	1.90	3.45
Rajhara	−50	62.54	2.59	7.30	65.20	1.97	3.14
Donimalai	−30	65.00	1.90	2.60	67.50	2.00	2.00
Bolani	−75	56.60	5.00	6.30	60.82	1.44	4.33
Noamundi Soft Laminated	−50	58.00	2.20	7.40	59.70	1.52	6.70
Noamundi Lateritised Soft	−50	58.70	1.90	6.90	60.10	1.26	5.98
Noamundi Hard	−50	64.00	1.86	3.30	66.70	1.28	2.20
Barsua Soft Laminated	−50	62.03	3.15	4.50	64.04	1.82	3.28
Barsua Lateritic	−50	58.87	1.48	7.04	60.40	1.06	6.20
Barsua Massive	−50	63.60	2.43	3.86	66.10	1.78	2.21

Results on heavy media separation carried out at specific gravity of 3.0 for washed -50 +10 mm lumps for different types and composite iron ore samples from Barsua mines are shown in Figure 1.2. The results are very well corroborated with the mineralogy of the samples. Studies were extended on heavy media separation of iron ore samples from different sources. The study indicates a clear improvement in the grade for the sink product particularly in respect of alumina. The floats can be treated for further recovery of iron values after size reduction. Presently in India,

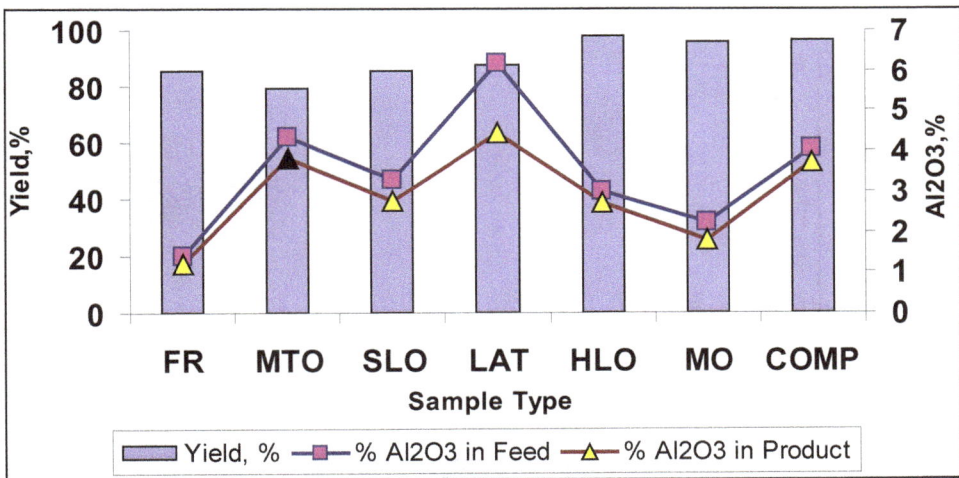

Figure 1.2: Results on Heavy Media Separation of Iron Ore Samples from Barsua Mines, SAIL

no plant is working with HMS in view of the availability of relatively better grade ores and for economic reasons. HMS will assume significance in utilising the lean resources. Figure 1.2 shows results on heavy media separation of different types of iron ore samples from Barsua mines, SAIL

Processing of Iron Ore Fines

Liberation of gangue from the ore minerals increases with decreasing size and the lateritic/aluminous material contained in -10 mm fraction produced from screening/washing, which could be rejected to some extent by gravity separation methods like jigging, tabling, spiralling and heavy media separation taking advantage of difference in specific gravity. Considering the difference in magnetic susceptibility of minerals percentage of alumina and silica can be lowered in the ore by magnetic separation. Degree of separation depends upon the mode of occurrence of the minerals in the ROM ore. Alternatively, hematite can be roasted in a reducing atmosphere to magnetite. The reduction roasting process can be followed by a low intensity magnetic separation. The results show that iron ore fines respond positively to reduction of alumina and silica by jigging and magnetic separation methods. Splitting of the feed and treatment of coarse fraction by jigging in combination with finer fraction by spiralling leads to improvement in results. Floatex density separator which works on the principle of a hindered settling classifier led to substantial reduction of alumina in 1.4 mm ore fines. Floatex density separator in tandem with a spiral is expected to give desired separation. Salient results on processing of a low grade ore with varying granulometry using a combination of hydrocyclone and a flowing film type gravity separator are shown in Figure 1.3. As it is evident, concentrate grade and iron recovery are very much influenced by increasing liberation with decreasing particle size and also the capability of the gravity separator for treating fine particles. Figure 1.4 shows the effect of magnetic field intensity on processing of an ore fines sample ground below 0.2 mm while Table 1.6 summarises the results of the studies carried out at

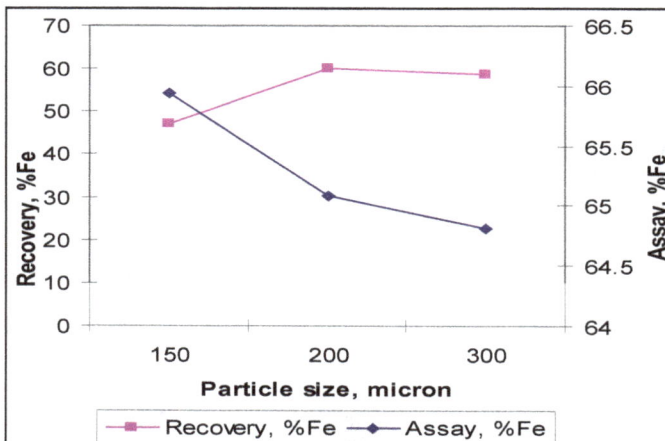

Figure 1.3: Effect of Granulometry on Concentration of Low Grade Iron Ore Using a Flowing Film Concentrator

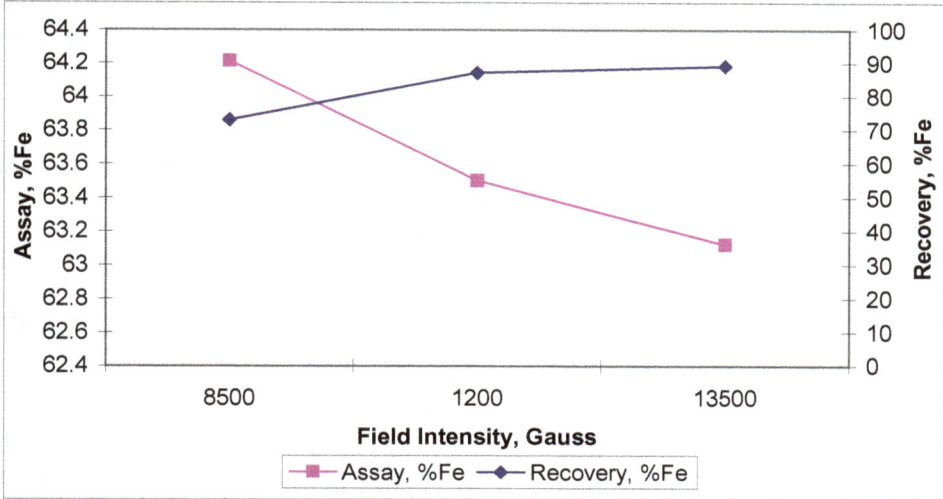

Figure 1.4: Effect of Magnetic Field Intensity on Concentration of Iron Ore Fines

CSIR-NML on concentration of iron ore fines involving different techniques. These results show that off grade iron ore fines can be upgraded and suitably utilised for iron making. A gravity-cum-magnetic separation process was developed for the processing of iron ore fines from Gandhamardan. It was possible to get a product assaying 66.75 Fe with an iron recovery of 61.9 per cent. Similarly, based on the

Table 1.6: Typical Results on Concentration of Iron Ore Fines

Source and Feed	Feed Assay, Per cent Fe	Process Techniques	Product Wt. Per cent	Product Assay, Per cent Fe
Noamundi washed fines	58.11	Splitting of feed followed by jigging of -9 +0.841 mm and tabling of -0.841 mm fraction	70.2	61.48
-3.36 mm classifier sand from Noamundi	57.40	Jigging	80.0	61.01
		Jigging of -3.36+0.6 mm and spiralling of -0.6 mm	63.2	62.25
Bonai,-3 mm	57.60	Jigging and tabling	51.6	61.31
Kudremukh	42.17	Magnetic separation and flotation	60.4	62.90
Kavuthimalai	34.60	Magnetic separation	50.1	63.15
Ongole	31.60	Reduction roasting and magnetic separation	39.6	65.50
Gua	59.00	Jigging	63.9	63.80
Dalli	58.52	Jigging and spiralling	65.8	65.40
Barsua	64.51	Spiralling	77.5	65.51

detailed characterisation, laboratory and pilot scale beneficiation studies, process was developed for processing of a low grade and fine grained ore to a product assaying over 65 per cent Fe, suitable as feed to DRI pellet plant.

Concentration of Iron Ore Slimes

The conventional practice of iron ore washing produces slimes (below 150 micron) which are not utilised and stored in tailings pond. Besides the loss of iron values, it poses environmental hazard. It is estimated that the generation of slimes is 10-25 per cent of the iron ore mined and it amounts to 18 million tonnes per year (Singh *et al.*, 2004). Iron ore slimes accumulated in different mines in the country alongwith their chemical characteristics are shown in Table 1.7 (Sengupta *et al.*, 1990). Any attempt to recover values from these slimes will eventually lead to additional supply of material for sinter feed and reduction of tonnage of slime to be discharged to tailing dams. However, proper characteristics need to be evaluated to find the amenability to beneficiation to produce a suitable additional raw material. R&D studies have been carried out on characterisation and beneficiation of slimes to recover iron values. Typical results are discussed below :

Table 1.7 : Production of Slimes by different Iron Ore Washing Plants

Mines	Quantity, MT	Assay, Per cent		
		Fe	Al_2O_3	SiO_2
Barsua	0.6	52.5	9.88	7.62
Bailadilla 5	0.5	61.2	2.81	6.84
Bailadilla 14	1.2	62.8	3.68	4.26
Bolani	0.4	59.8	4.8	4.10
Daitari	0.3	60.0	4.52	2.30
Donimalai	1.0	57.9	6.28	6.42
Kiriburu	1.6	60.4	4.96	2.96
Kudremukh	15.0	26.6	1.82	51.02
Noamundi	0.45	55.0	7.8	5.02

Hydrocycloning and Gravity Separation

Characterisation of the iron ore slimes samples indicated that in several cases there is a preferential concentration of gangue minerals in the finer fractions. There exists possibility of preferential rejection of gangue mineral particles using hydrocyclone. Physical and chemical properties of selected slimes sample are presented in Table 1.8. Results of the studies on processing of iron ore slimes using hydrocyclone are summarised in Table 1.9.

It is often found that slimes rich in laterites/limonites do not respond well to hydrocyclones or the conventional gravity separators in lowering the alumina to 2 per cent or below. Under such cases hydrocloning, in tandem with a suitable gravity separation device capable of treating fine particles, proves helpful in reducing alumina in the slimes. Spirals of recent origin with improved profiles are able to provide better retention of ultrafines and lowering of losses in the fine size ranges.

Table 1.8: Physical and Chemical Characteristics of Iron Ore Slimes

Plants	d_{80}, Micron	Assay, per cent			
		Fe	Al_2O_3	SiO_2	LOI
Barsua	50	49.1	11.4	9.1	9.3
Bolani	250	57.5	6.7	5.8	4.3
Kiriburu	105	54.1	8.1	6.6	8.2
Noamundi	75	57.8	8.3	4.0	5.2

Table 1.9: Results on Concentration of Iron Ore Slimes by Hydrocycloning

Source	Assay of Feed, per cent			Product	
	Fe	SiO_2	Al_2O_3	Wt. Oer cent	Assay, Per cent Fe
Noamundi	58.50	4 22	5.90	66.4	62.10
Bolani	49.80	6.60	8.10	49.7	58.00
Meghataburu	57.65	4.38	5.95	54.9	63.52
Joda (Flaky)	48.72	11.23	12.27	45.1	61.76
Joda (Hard)	43.80	11.65	14.25	40.7	49.98
Gandhamardan	54.80	7.90	6.10	64.5	60.50
Bellary Hospet	57.60	8.00	5.90	60.0	66.90

Use of Centrifugal Separators

Iron ore slimes consisting of particles below 150 micron do not respond to conventional gravity separation equipment. But some of the recent fine gravity separators, in particular centrifugal separators, like multi-gravity separator (MGS), Kelsey jig, Falcon separator and Knelson concentrator have got potential for recovering mineral values from fines. By using MGS it has been possible to reduce alumina content in typical iron ore slimes samples to 2 per cent (Singh *et al.*, 2000). MGS basically uses a high 'g' forces to bring about separation in the fine size ranges. Conceptually it can be visualised by rolling the deck of a conventional shaking table into a drum and then rotating it. Typical results obtained using MGS are shown in Figure 1.5. However, the recent studies carried out at CSIR-NML on superconcentration of iron ore slimes from Hospet indicated that so far as the yield and reduction of silica are concerned, the results obtained by WHIMS are at par with those obtained by MGS.

Magnetic Separation (Low and high intensity with wet and dry type)

Depending upon the nature of iron bearing minerals, a low or high intensity magnetic separator or a combination of the two can be used for processing of a given iron ore. Particularly for the recovery of magnetite a low intensity magnetic

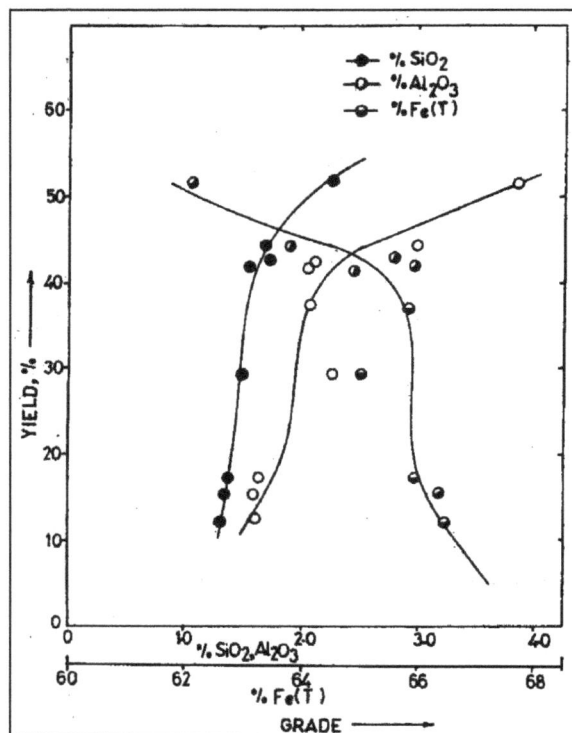

Figure 1.5: Results on Reduction of Alumina and Silica using MGS

separator with a magnetic field intensity of 1000-2000 gauss in the concentration zone is considered suitable. For the weakly magnetic minerals like hematite and limonite a high intensity separator with an intensity of 7000-20,000 gauss is used.

Several studies have been carried out at CSIR-NML and in other laboratories in India for processing of iron ore slimes from different sources by this technique. The results show that it is possible to reduce alumina to the level of 2-3 per cent in the product. In the studies undertaken at CSIR-NML, on a finely disseminated siliceous iron ore from Ukraine, it has been possible to upgrade the sample (40.75 per cent Fe) to a product assaying 59.04 per cent Fe with a yield of 61.8 per cent, by two stages of magnetic separation.

Flotation and Selective Flocculation

Concentration of iron ores by froth flotation is an established process (Houot, 1983; Iwasaki, 1983). Iron bearing minerals like hematite and goethite can be floated using standard collectors like sulphonate, oleate, sulphate. Conversely, silica can be floated using anionic or cationic reagents. Organic reagents like starch or gum are used as depressant for iron oxides. Selective flocculation is another emerging technique for separation of minerals in the fine size ranges. In this case separation is achieved by selective dispersion followed by flocculation of the iron bearing minerals using a suitable organic polymer. Considerable work has been carried out at CSIR-NML on superconcentration of iron ore fines by reverse flotation of silicate minerals and salient results are discussed in the subsequent section. Reverse flotation of silica proved to be very effective in enriching iron content to 66.46 per cent Fe in the magnetic concentrate produced from a fine grained siliceous iron ore sample from Ukraine (Singh *et al.*, 2004). As mentioned above, the sample contained 40.75 per cent Fe with 39.42 per cent SiO_2 The effort on development of selective flocculation –flotation process has not shown any breakthrough for beneficiation of Indian iron ore slimes. The lack of selectivity of the standard flocculants like starch and polyacrylamide for iron oxide-alumina system is attributed to their similar crystal structures and the chelation characteristics of constituent metal ions. It needs further investigation to develop selective flocculation technique for recovery of iron oxide minerals from Indian iron ore slimes.

Column Flotation for Beneficiation of Indian Iron Ore

CSIR-NML has been involved in carrying out R&D studies in column flotation technology. The flotation column (Figure 1.6) developed by CSIR-NML has been extensively used for studying beneficiation of various low grade ores including iron ore samples from Kudremukh and Goa (Prabhakar *et al.*, 1997). The Kudremukh Iron Ore Processing Plant was commissioned with a treatment capacity of 22.6 million tonnes of iron ore per year for production of high quality blast furnace and direct reduction grade pellets for export. But the silica content in the final iron ore concentrate was found to be high and the recovery of mineral was also not satisfactory. Based on the detailed studies undertaken on different samples it was possible to get a concentrate assaying SiO_2 <2.5 per cent with over 95 per cent iron recovery by column flotation technique. Column flotation work has also been extended to the processing of iron ores of Goa origin for M/s Sociedadel De Fomento Industries Limited, Goa. The present beneficiation circuit consisting of gravity and magnetic separation units is able to produce iron ore concentrate assaying 65-66 per cent Fe. A detailed laboratory study and in-plant column flotation trials conducted on the concentrate from HGMS have shown that product analysing 67 per cent Fe and around 2 per cent SiO_2 and Al_2O_3 with iron recovery of 85-90 per cent can be obtained by single stage column flotation. Recently studies have been extended for beneficiation of slimes from hematitic ores. The results on pilot scale column flotation of ore slimes from Jindal Steel Works has been very encouraging. Thus column flotation technology offers an attractive option for producing high grade iron ore concentrate from low grade ores, fines and slimes.

Sintering of Iron Ore Fines and Utilisation of Blue Dust for Alumina Reduction

Sintering is an effective method of utilising iron ore fines. Use of quality sinter in a blast furnace provides overall economy in the iron making process. Studies have been undertaken on sintering of iron ore fines at CSIR-NML since early sixties to cater to the need of the iron and steel industry in the country. Pot grate serves as a cross section of the actual sinter strand and in laboratory, pot grate studies are helpful in finding suitability of the raw materials, effect of different variables on physical, chemical and physico-chemical properties of sinter and to optimise

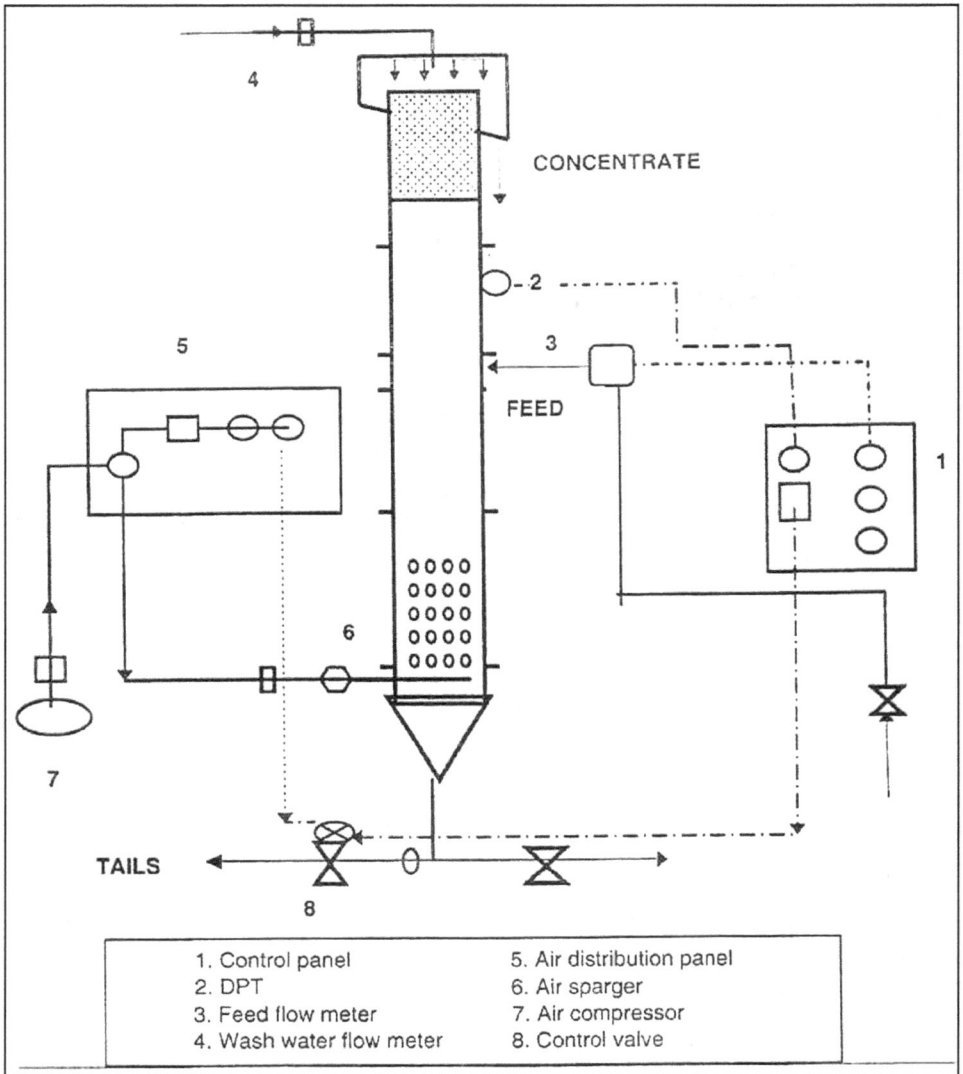

CONCENTRATE

FEED

TAILS

1. Control panel	5. Air distribution panel
2. DPT	6. Air sparger
3. Feed flow meter	7. Air compressor
4. Wash water flow meter	8. Control valve

Figure 1.6: Schematic of CSIR-NML Pilot Scale Column Flotation Unit

the parameters. The finding of the pot grate studies are subsequently simulated in pilot plant or actual plant practice. The capacity of sinter plants in India is shown in Table 1.10 (Yadav, 1997).

With the mechanisation of mines, the generation of fines has been increasing. Besides, natural occurrence of super-fine blue dust having high iron and low alumina content needs avenues for its effective utilisation in blast furnace through sinter. Such as blue dust is an ideal agent for reducing the alumina load in the burden. In spite of being good grade, both the cyclone underflow from conventional iron ore washing plants and blue dust are extremely fine (25-50 per cent below 74 micron) and thus have found only limited use in sintering, though considerable amount of research work has been carried out for utilisation of these and other ultrafines.

Table 1.10: Sinter Plants in India

Steel Plant	Size of the Machine, m	Capacity, MT/yr	Percent Sinter	Basicity
Bhilai	4x50	2.04	56	2.4-2.5
	3x75	3.14		
	1x80			
	1x312	3.20		
Bokaro	3x252	6.90	68	1.86
		4.10		
Durgapur	2x140	1.50	43	1.81
	1x180	1.60		
Rourkela	2x125	1.50	54	2.05
	1x192	1.57		
Tata Steel	2x75	1.50	60	2.36
	1x192	1.61		
Vizag	2x312	7.9	90	3.4

Studies, undertaken of CSIR-NML, for effective utilisation of blue dust encountered at the iron ore mines, have focussed on the extent to which it can be successfully mixed with iron ore fines without affecting the metallurgical quality of the sinter and its productivity. The comprehensive study for utilisation of blue dust (51.6 per cent -150 micron) to the extent of 10-50 per cent in sintering were undertaken while working with Noamundi washed iron ore fines (3.6 per cent -150 micron). Results on sintering of Noamundi iron ore fines with blue dust are shown in Table 1.11. It is seen that even with 50 per cent blue dust, sinters with shatter size stability of 63.8 per cent could be achieved. In another study on sintering with iron ore from Kiriburu mines of M/s NMDC Ltd., the amount of blue dust in the mix was varied from 10-50 per cent with respect to iron ore fines keeping the basicity ratio fixed at 1.6. Results indicated that best sinter was obtained with 30 per cent blue dust in iron ore fines. The basicity for these experiments were maintained at 2 with 25 per cent return fines (Maulik, 1985).

The primary objective of the comprehensive sintering study, undertaken on iron ore fines and blue dust from Bailadila deposit for M/s National Mineral Development Corporation, Hyderabad, was to produce a sinter having a productivity of at least 1.27 t/hr/m². The proportions of ingredients of the raw mix compositions were calculated based on a software developed at CSIR-NML. Variables studied included moisture, coke breeze, return fines, blue dust and basicity of the sinter. It was seen that from a basicity of 2.0 and onwards, sinters with acceptable strength could be produced. Maintaining a bed height of 400 mm, sinters with Tumbler Index of about 65-66 and 71-72 could be produced at basicities of 2.0 and 2.25, respectively. This study also indicated that good sinters could be produced with addition of blue dust upto 50 per cent in iron ore fines. In this work, regression analysis between strength and basicity indicated that there existed a linear relationship between them, and strength increased with increasing basicity. It was also noted that the strength increased with the increasing fines content. However, an optimum quantity of blue dust addition showing a maximum strength was indicated. It was also seen that with higher basicity, higher amount of blue dust additions could be made.

Table 1.11: Results of Sintering of Noamundi Iron Ore Fines using Blue Dust

Sl.No.	Water, Per cent	BD in Ore Fines, Per cent	Coke Breeze, Per cent	Speed of Sintering, cm/min.	Yield, T/hr/m²	R/F Produced, per cent	Shatter Sixe Stability, Per cent
1.	6	10	4.0	2.31	1.44	24.3	62.8
2.	6	20	4.0	1.72	1.41	23.1	58.3
3.	6	25	4.0	2.11	1.29	25.3	64.4
4.	6	33	4.0	2.03	1.28	25.1	59.4
5.	6	50	4.0	1.65	1.01	29.9	63.8
6.	9.5	100	4.0	1.45	0.80	35.7	50.7
7.	7	33	4.0	2.11	1.37	21.0	59.5
8.	8	33	4.0	2.21	1.41	22.7	59.0
9.	7	50	4.0	2.03	1.25	25.8	58.4
10.	8	50	4.0	2.41	1.74	32.9	47.6

Several studies as discussed above, clearly indicate the possibility of using the ultrafines recovered from processing of iron ore fines and slimes, thereby not only reducing the alumina level but also leading to conservation of resources and minimising pollution.

Pellitization of Beneficiated Fines and Slimes

Pelletizing is the process of converting iron ore fines into "uniformed sized iron ore pellets" that can be charged into the blast furnaces or for production of Direct Reduced Iron (DRI). Pellet plants had an added advantage that they could be stand alone units established close to the mines and the pellets transported to the steel plants, whereas, sinter plants needed to be necessarily set up within steel plant premises as sinter could not be transported over long distances without deterioration in quality. They are almost pollution free, generating no solid or liquid residues.

In the face of shrinking world reserves of high grade ores, pellets from low grade ores is one of the best options due to their excellent physical and metallurgical properties. Moreover, due to their high strength and suitability for storage, pellets can easily be transported over long distances. The major benefits of pellets are as follows :

☆ Standardization : uniform size range, generally within a range of 9-16 mm

☆ Purity : 63-68 per cent iron, mainly Fe_2O_3

☆ Cost effectiveness : virtually no loss on ignition

☆ Porosity : uniform porosity of 25-30 per cent allows fast reaction

☆ Metallization : high metallization, > 90 per cent

☆ Strength : high and uniform mechanical strength

☆ Transportable : very low degradation during transportation

Low grade ores and fines can suitably be utilised by their enrichment to the required specification followed by pelletization of the fine grained concentrate. Pellets are made by mixing finely ground concentrate with a small quantity of binder (usually bentonite or hydrated lime) and balling the mixture in rotating drums or saucer like discs. It is important to control moisture content of the mix. The green pellets usually 10-13 mm in diameter, are then hardened in gas or oil fired furnaces or kilns at about 1300°C. Most pelletizing systems in use are of the 'straight grate' or 'grate-kiln' type but vertical shaft furnaces are also used. Pelletization on magnetite and hematitic iron ore samples were carried out with a view to establish the parameters for production of direct reduction/blast furnace quality pellets. Detailed studies carried on Kudremukh iron ore concentrate resulted in recommendation of various parameters such as grind level of 1600 blaine and a basicity level of 0.8 using hydrated lime as both flux and binder. The pellet produced under the above conditions had a green compressive strength of 07-08 kg/pellet, drop strength from 450 mm of 3-4 drops, moisture content of 6-7 per cent and cold crushing strength after heat hardening at 1275°C of 300-460 kg/pellet. The pellets had reducibility of 56-57 per cent porosity of 24-25 per cent. Swelling index of 16-17 per cent and Tumbler and Abrassive indices of 93-94 per cent and 6-7 per cent, respectively. At the instance of M/s Ashapura Industries, Ltd, Mumbai studies were undertaken to examine suitability of bentonite from different sources as binder for iron ore samples of Indian and foreign origins. In view of the high energy cost in heat hardening of pellets, technology has been developed on micro-pelletization and sintering. The pellet grade fine concentrate can be converted to micro-pellets which do not need heat hardening and can be used as partial feed to sintering alongwith normal sinter fines. This also leads to marked improvement in sinter productivity (Singh, 2014).

Conclusions

India is endowed with large reserves of iron ore which is the basic raw material for iron and steel making. But Indian iron ores are characterised by high alumina and the high grade lumpy ores are getting depleted. Beneficiation of iron ores for reduction of alumina is imperative for efficient operation of blast furnace. Due to the complex nature of the association of iron and alumina bearing minerals, the conventional iron ore washing schemes have limited success in reducing alumina to the desired level in low grade ores. Considering the increasing demand of quality iron ores and limitations of conventional beneficiation circuits, there has been need to adopt improved technology for beneficiation of low grade ores. In the light of recent developments and the studies carried out, innovative process technologies have been developed for processing of low and lean grade iron ores, waste dump fines and slimes for producing quality iron bearing raw material for iron and steel making. In view of the increasing demand, depleting high grade resource and conservation there has been increasing trend for adoption of newer technologies for processing of iron ores in India.

Acknowledgements

The author wish to thank Dr. S. Srikanth, Director, CSIR-National Metallurgical Laboratory, Jamshedpur for kindly permitting to submit this paper for publication. Thanks are also due to sponsors for financing the research projects and providing iron ore samples for the studies.

References

1. Anonymous, Indian Minerals Year Book: 2011, Indian Bureau of Mines, Nagpur, 2012, p.47.

2. Houot, R., Int. J. Mineral Processing, 1983, 10, p.183.

3. Iwasaki, I., Mining Engineering, 1983, 35, p.622.

4. Maulik, S.C., 1985, CSIR-NML Investigation Report No. 1209/85, Jamshedpur, India.

5. Prabhakar, S., Bhaskar Raju, G., Subba Rao, S. and Sankaran, C., In : Processing of Fines, CSIR-NML Jamshedpur Publication, 1997, p.103.

6. Pradip, Metals Materials And Processes, 1995, 6 (3), p.179.

7. Rao, P.V.T., Sripriya, R. Murthy, V.G.K., The Indian Mining and Engineering Journal, April 1995, p.1.

8. Sahu, N.C., The Indian Mining and Engineering Journal, April 1995, p. 9.

9. Dey, Shobhana, Pany, Santosh, Mohanta, M. K. and Singh, R, "Utilization of iron ore slimes : A future perspective", Separation Science and Technology, 2012, 47 (5), 769-776.

10. Sengupta P.K. Prasad N., In: Procd. Iron Ore Processing and Blast Furnace Iron Making, IBH and Oxford Publication, New Delhi, 1990, p.8.

11. Singh, R., 2014, CSIR-NML Investigation Report, Jamshedpur, India.

12. Singh R., Bhattacharyya P.K., Bhattacharyya K.K. and Maulik S.C., Metals Materials And Processes,Vol.4, No.2 and 3, 2004, p.157.

13. Singh, R., Bhattacharyya, K.K., Bhattacharyya, P. K. and Maulik, S.C., In : Research Techniques in Mineral Processing Waste and Environment Management, Allied Publishers, New Delhi, 2000, p.143.

14. Singh, R, Rath, R.K., Nayak, B., and Bhattacharyya, K.K. "Development of process for beneficiation of low grade iron ore samples from Orissa, India", In: Proceedings of XXV International Mineral Processing Congress, Brisbane, Australia, Sept. 2010.

15. Weiss, N.L., SME Handbook of Mineral Processing, SME/AIME Publication, New York, USA, 1985, 20-2.

16. Yadav U.S., Pandey B.D., Elijgh N., Sriram S. and Ghosh S., Tata Search, 1997, p.11.

Chapter 2

Value Addition and Beneficiation: Developing a Capabilities Driven Beneficiation Framework for the Iron and Steel Industry in Zimbabwe

Elias Matinde[1] & Ephraim Makoni[2]

[1]Scientific and Industrial Research and Development Centre,
Harare, Zimbabwe
E–mail: emzim2004@yahoo.com
[2]Graduate School of Management,
University of Zimbabwe, Harare, Zimbabwe

ABSTRACT

Purpose: This paper highlighted the critical technical, innovation, managerial and organizational capability gaps required to develop a capability-driven beneficiation framework for the iron and steel industry in Zimbabwe.

Methodology and Approach: This study adopted a case study based approach to analyze the current technological and innovation capabilities, and the key technological and innovation trends and drivers within the iron and steel industry in Zimbabwe. The survey instrument was developed based on a hybrid of the OECD-CIS harmonized methodology, the Boston Consulting Group, and the Price Waterhouse Coopers methodologies for national innovation surveys.

Findings: The results show that only a marginal number of the companies in the iron and steel sector in Zimbabwe possessed the requisite technological and innovation capabilities necessary to drive competitiveness. In addition, the study revealed critical value chain gaps within the iron and steel sector, and the constraints to the full value chain utilization. On the positive note, the results show that the companies were adopting balanced measurement metrics to benchmark their technological and innovation capabilities.

Practical implications: The paper evaluated the technical, managerial and organizational capabilities based on global benchmarks. In depth evaluation of the technological and innovation trends and drivers, the capabilities measurement metrics, and the critical success factors to the technological and innovation management strategies within the iron and steel sector in Zimbabwe were carried out. Finally, the paper made recommendations to integrate the sector specific technological and innovation best practices into the broader national beneficiation framework.

Practical limitations: The paper focused on a specific industry sector in Zimbabwe. However, the problems in local iron and steel sector resonates the economy based wide problems currently being experienced in the country. As a result, any future research needs to focus on the dynamic interaction between the interlinked business sectors in the broader Zimbabwean economy.

Originality and value: This article provides a detailed gap analysis of the critical success factors imperative to building the technological and innovation capabilities.

Paper type: Survey-based research paper

Keywords: *Beneficiation/value added processing, Technological and innovation capabilities, Innovation management, Incremental innovation, Radical innovation, Breakthrough innovation, Capacity utilization, Value chain maximization, Beneficiation framework.*

Introduction

Beneficiation of minerals and energy resources is a buzz word in Zimbabwe at the moment. In general, beneficiation or value added processing is defined as the processes and steps taken to transform primary commodities into an intermediate or finished product with higher value than otherwise when exported in its raw form (Baxter, 2005). Beneficiation and value addition of mineral resources not only increase the export value of primary commodities, but also creates knock-on effects within the wider economy through employment creation, linkages, knowledge and capacity building in upstream industries, and the industrialization of the whole economy (Baxter, 2005). The value addition of mineral resources before export was prioritized in the ZIM ASSET economic blue print document as a tool to transform the country's comparative advantage into a sustainable competitive advantage (Government of Zimbabwe, 2014). The ZIM ASSET economic blueprint was promulgated at a time when the local iron and steel sector was facing severe viability issues. As a result, specific emphasis and priority was given towards resuscitating the iron and steel sector and ensuring that the country maximizes the value across the whole iron and steel value chain (Government of Zimbabwe, 2014).

Literature Survey and Theoretical Framework

Innovation is broadly defined as the successful entry of a new science or technology based product or process into a particular market (Branscomb, 2002). Innovation can further be defined as a process of implementing new ideas, creating dynamic products or improving existing services, and thus, can be a strong catalyst for growth and success in a constantly changing business environment (Branscomb, 2002). Technological innovation consists of the capacity to develop new products satisfying the current market needs, apply appropriate technologies to produce new products, to develop and adopt new product and process technologies to satisfy future needs, and, the capacity to respond to accidental technology activities and unexpected opportunities created by competitors (Adler, 1991; Lawson, 2001). The concept of technological innovation capabilities further refers to the ability to continuously transform knowledge and ideas into new products, processes and systems for the benefit of the firm and its stakeholders (Adler, 1991; Lawson, 2001). Technological capability can also be broadly defined as the knowledge and skills required for firms to choose, install, operate, maintain, adapt and develop technologies, and necessitate purposive efforts aimed at assimilating, adapting and modifying existing technologies and/or developing new technologies to meet existing and or future market needs (Adler, 1991; Malerba, 1992; Lall, 1992; OECD, 2005)

Pioneering studies on the role of technological change and innovations as a source of economic change, and catalyst for business cycles, date back to the seminal works by Joseph Schumpeter (Schumpeter, 1934). According to Schumpeter, innovation consists of the introduction of new products, new methods of production, developing and opening up new markets, development of new sources of supplies of raw materials or other inputs, and the creation of new market structures or new form of an organization in an industry (Schumpeter, The Theory of Economic Development: An Inquiry into Profits, Capital, Credit, Interest, and the Business Cycle, 1934; Schumpeter, Business Cycles: A Theoretical, Historical, and Statistical Analysis of the Capitalist Process, 1939). Schumpeter further defined technological innovation as a new combination of means of production, that is, as a change in the factors of production (inputs) to produce output products (Schumpeter, 1939). Schumpeter argued that economic development is driven by innovation through a dynamic process in which new technologies replace the old via "creative destruction" (OECD, 2005; Schumpeter, The Theory of Economic Development: An Inquiry into Profits, Capital, Credit,Interest, and the Business Cycle, 1934). According to Schumpeter, it is the "radical and break through innovations" that create major disruptive changes, as opposed to "incremental innovations" which continuously and incrementally advance the process of change (Schumpeter, The Theory of Economic Development: An Inquiry into Profits, Capital, Credit,Interest, and the Business Cycle, 1934; OECD, 2005).

From the seminal studies, Schumpeter proposed five types of innovations, namely, the introduction of new products, introduction of new methods of production, opening of new markets, development of new sources of supply for raw materials or other inputs, and the creation of new market structures or business models in an industry (Schumpeter, 1934; Schumpeter, 1939). Although modern innovation management theories have since evolved, they are still strongly embedded in the original Schumpeter's views where ultimate reason to innovate is to improve firm performance (OECD, 2005). In addition to studies by Joseph Schumpeter (Schumpeter, 1912; Schumpeter, 1939), the recent studies also provide overwhelming evidence that innovation enhances market position and economic performance for the innovating company (OECD, 2005; Wagner, 2013). According to the OECD studies, a new product innovation often leads to market advantage for the innovator, while a process innovation leads to cost advantages over competitors, hence, "allowing a higher mark-up at the prevailing market price", and be able to use the "combination of lower price and higher mark up than competitors to gain market share and increase profits" (OECD, 2005; Wagner, 2013). In addition, "companies can also increase the demand through product differentiation, targeting new markets and by influencing demand for existing products" (OECD, 2005). The impact of product innovations on firm performance therefore ranges from the "effects on sales and market share of products to changes in productivity and efficiency", and "the outcome of such product innovations can be measured by the percentage of sales derived from new or improved products" (OECD, 2005).

The Boston Consulting Group publishes global annual survey results of the most innovative companies in 2013 based on how the respondents ranked their innovation performance relative to their peers in the market place (Wagner, 2013). From the study findings, the top fifty innovative companies outperformed their rivals in terms of weighted shareholder returns, revenue growth and margin growth (Wagner, 2013). According to these authors, innovating companies continue to create value in the long term by incorporating innovations in their corporate DNA, with success factors being determined by the rate at which such innovations and innovative ideas are reduced to practice (KPMG, 2012; Wagner, 2013).

Why Acquire Technological and Innovation Capabilities

The strategic importance of acquiring technological and innovation capabilities is widely accepted within management circles as a tool for building sustainable competitive advantage in organizations (Chiesa, 1996; OECD, 2005; Kaplan, undated; Gerybadze, 2010; Saunila, 2013; Wagner, 2013). For example, research and development (R&D) and innovation have long been acknowledged as the drivers of change and the key determinants of growth across many industries and service sectors (Chiesa, 1996; OECD, 2005; Gerybadze, 2010; Saunila, 2013). According to these authors, organizations that continuously invest in R&D, and persistently expand their base of technological capabilities often attain stable growth and strong financial performance (Gerybadze, 2010; Saunila, 2013). Based on the cycle of innovation, the high margins and above-average returns attained by innovative organizations are then ploughed back to be reinvested in more R&D than rivals, and are therefore such organizations are able to manage new product development pipeline in perpetuity, (Gerybadze, 2010).

The Boston Consulting Group survey identified the key attributes that separate strong innovators from their weaker counterparts (Wagner, 2013). According to this study, the strong innovators have top management that is committed to innovation as a competitive advantage, leverage on intellectual property (IP) and manage a portfolio of innovative initiatives, leveraged on strong customer focus, and insist on strong processes as a driver of strong performance (Wagner, 2013). When complimented with other factors such as the corporate strategic direction, organizational capabilities and dynamic interaction with the firms' ecosystem, the innovating organizations are therefore able to build dynamic competitive capabilities based on strong managerial capabilities at both the corporate as well as the business unit level (Gerybadze, 2010). Such organizations have innovation systems that are strongly supported by their corporate strategies to develop effective innovation routines, strong new product pipeline, and are able to develop the ability to evaluate and absorb risks better than rivals leading to the deployment of investment policies that support expensive and often risky R&D projects (Gerybadze, 2010).

Taxonomy and Innovation Framework in Developing Countries

A number of exogenous systematic factors impact the taxonomy of the technological and innovation landscape in developing countries, for example, macroeconomic uncertainty, inadequate physical infrastructure, institutional fragility, lack of social awareness on innovation, barriers to business start-ups, and lack of public instruments for business support and management training (OECD, 2005; Pietrobelli, 2006; UNCTAD, 2007). Most companies in the developing countries operate in unusual economic and innovation environments due to the existence of state owned enterprises and/or parastatals, and the lack of competition discourages innovation or distorts local markets of innovative potential (OECD, 2005). For most developing countries, competitiveness is mostly based on the exploitation of natural resources and/or cheap labour rather than on efficiency and differentiated products, leading to informal organization of innovation and fewer R&D projects (OECD, 2005).

Generally, fewer resources are devoted to system-wide technical and innovation activities in developing economies, thereby reducing the innovation potential of enterprises (OECD, 2005; Pietrobelli, 2006). In most cases, the state is the major player in R&D funding and execution, mainly due to the low level of resources devoted to R&D from the private sector (OECD, 2005). In addition, the flow of information pertaining to innovation within national innovation systems is fragmented, and is characterized by weak or complete absence of linkages between public scientific institutions and the manufacturing enterprises (OECD, 2005). Therefore, the weak or absent linkages inhibit the capacity of firms to overcome technological related problems, and hence drives the firms towards solutions that rely mostly on the acquisition of embodied technology (OECD, 2005; UNCTAD, 2007). In most cases, these links between the manufacturing companies and the science and technology institutes in most developing countries are weak because of poor working relationships and understanding between manufacturing companies and these institutions due to the misplaced perceptions on the public research institutes' abilities to carry out meaningful industrial research (Lall, 2004; Pietrobelli, 2006).

Measuring Technological and Innovation Capabilities

The most compelling reason to measure technological and innovation capabilities is to develop the technological and innovation capabilities of the innovating organization. The capability view of the firm assesses the extent to which the company's competences, culture and conditions support the conversion of innovation resources into opportunities for business renewal (Muller, 2005).

As a result, a holistic innovation measurement system becomes an enabler and catalyst for organizational diagnosis by unearthing the specific weaknesses in the company's innovation capabilities through on-going innovation performance audits, and through focusing employees' attention towards the right projects, activities and behaviors from an innovation performance point of view (Kaplan, undated). In this case, developing a holistic innovation measurement system assists in employee motivation through right goals and the incentives to drive employee involvement, thereby ultimately increasing the innovation efficiency and effectiveness of the innovating companies (Kaplan, undated).

Measuring technological and innovation capabilities also enables organizations to optimally allocate resources towards technical and managerial activities that continuously generate dynamic capabilities (Muller, 2005; Gerybadze, 2010; Saunila, 2013; OECD, 2005). Such decisions are predicated on the availability of inimitable combinations of resources that cut across all innovation functions such as R&D, product and process development, manufacturing, human resources and organizational learning (Lawson and Samson, 2001).

Despite the widely accepted paradigm on the importance of measuring technological and innovation capabilities, limited literature on the subject is available in the public domain, particularly for the developing countries. Innovation as a concept is also poorly understood in most developing economies, mostly due to its nature as a complex, dynamic and non-linear activity which makes its measurement a challenging and continuous learning process (Human Resources Research Council, 2008; OECD, 2005). Due to the complexity of the new approaches to innovation management, innovation is no longer regarded as an outcome resulting only from the performance of R&D, but is also taken to include results from a variety of non-R&D activities and expenditures (OECD, 2005; Human Resources Research Council, 2008). In most cases, activities leading to innovation may include the embodied technology through the acquisition of machinery, equipment, software and knowledge from outside the innovating organization (Chesbrough, 2003; OECD, 2005; Human Resources Research Council, 2008; UNCTAD, 2007). These views militate against adopting the technological and innovation capability models applied in advanced economies to developing economies at face value because of the economic taxonomies that characterize most of the developing countries (OECD, 2005).

Conceptual Framework for Technological and Innovation Capabilities

The importance of technological and innovation capabilities-driven economic development was highlighted in several studies, and has received wide attention at both the national and global levels. In Zimbabwe, the Government of Zimbabwe, through the Industrial Development Policy (Ministry of Industry and Commerce, 2011), the Science, Technology and Innovation Policy (Ministry of Science and Technology, 2012), and the Zimbabwe Nanotechnology Statement (Ministry of Science and Technology Development, 2012), has prioritized science, technology and innovation as the key drivers towards achieving a competitive and vibrant industrial growth through the development of new and competitive processes and products.

Gerybadze (2010) proposed a conceptual framework on how organizations can transform the dynamic capabilities into a sustained competitive advantage (Gerybadze, 2010). Figure 2.1 proposes a conceptual framework for developing dynamic and sustained technological and innovation capabilities adopted in this study (Gerybadze, 2010).

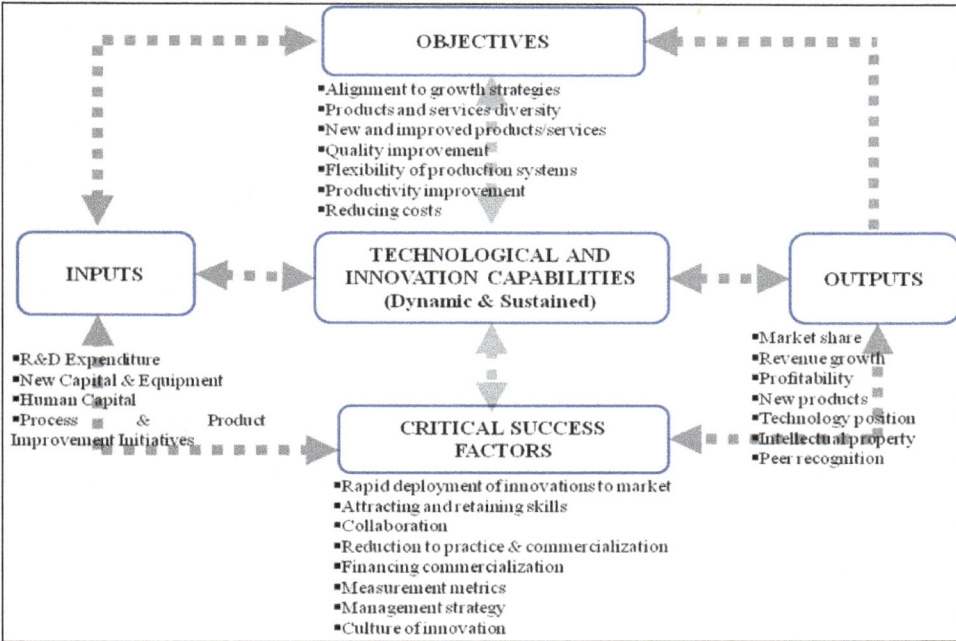

Figure 2.1: Conceptual Framework for Developing Dynamic and Sustained Capabilities (Gerybadze, 2010)

Overview of the Iron and Steel Value Chain

The iron and steel sector forms the back bone of economic development in any society, and has been credited for the industrial transformation of modern society (Manning, 2001). The iron and steel products form the basis for value added manufacturing and economic development, especially in the specialized sectors such as infrastructure, mining, construction, automotive, chemicals and agriculture.

The iron and steel value chain consists of distinctive but closely interlinked stages, namely, exploration, mining beneficiation, metallurgical processing (smelting and refining) and shaping, conversion, fabrication and assembly, manufacturing and end-use, and retail and distribution (Kumba Iron Ore, 2011). The metallurgical beneficiation and shaping typically involves smelting to convert the iron ore into pig iron, and shaping it in rolling mills into steel products (Kumba Iron Ore, 2011). The converters and fabricators then further convert the standard steel products into intermediate finished products, while the manufacturers and end-users consume both the standard steel products and intermediate products from converters (Kumba

Iron Ore, 2011). The key ingredients in the iron and steel and the stainless steel production include iron ore, fluxes (limestone and dolomite), metallurgical coke, ferrochrome, and nickel, with ferrochrome and nickel being the most critical in the production of stainless steel (Manning, 2001; Kumba Iron Ore, 2011).

Stainless steel represents one of the most value added products in the iron and steel value chain due to its indispensable applications in specialized sectors such as automotive, chemicals, energy, agriculture, mining, and manufacturing. Being a high value added product, the competitiveness of any iron and steel operation is therefore predicated on the technological capabilities to refine the high end stainless steel products for the various applications within the broader national economy.

Although there are many process options for the smelting and refining of steel, the individual choice is influenced by the raw material availability, final product characteristics and use requirements, and the capital economics (Total Materia, 2008). The critical raw materials to stainless steel process include iron, chromium, and nickel, and in some instances, molybdenum, tantalum, beryllium, and niobium are added depending on the steel grades (Total Materia, 2008).

Global Stainless Steel Dynamics

The global stainless steel production averaged 35 million tons in the period 2012, with the estimated compounded annual growth averaging 5.6 per cent per annum (International Stainless Steel Forum, 2013). In addition, the real demand for stainless steel was expected to surpass 5 per cent for the period 2012-2014 (International Stainless Steel Forum, 2013), with the buoyant demand in stainless steel products being driven by the demand in rapidly developing economies (Elliot, 2013). The continued growth and long term demand in stainless steel products remains bullish, driven by global mega-trends in specialized end uses in infrastructure, transportation, automotive, chemicals, petrochemical and energy sectors (Outokumpu, 2012). As a result, the sustained global growth provides real growth opportunities for both primary and specialized producers within the stainless steel sectors (Outokumpu, 2012). In response to this projected buoyant growth in demand, the global producers of steel and stainless steel products have prioritized the high end and specialty steel products for various applications (Wortler, 2010). As a result, building the technological and innovation competitiveness through continuous research and development was prioritized by the top management of most global producers of steel and stainless steel products (Wortler, 2010).

Overview of the Zimbabwean Iron and Steel Sector

The iron and steel industry in Zimbabwe traditionally applied the integrated blast furnace and the basic oxygen furnace processes at the Zimbabwe Iron and Steel Company (now New Zim Steel) works in Redcliff. Other key primary producers in the iron and steel production process include the Hwange Colliery Company (HCC) for metallurgical coke and coke products, the key ingredients in the production of iron and steel. Secondary producers include smelters and foundries that use the intermediate products from these primary producers to produce a variety of iron and steel products. It is clear that the iron and steel sector in Zimbabwe is characterized by incomplete value chain participation and poor value-in-use of steel intermediate products. Currently, the iron and steel sector in Zimbabwe is mostly focused on producing generic steel products, with little or no emphasis on producing more value added products such as stainless steel and high specialty steel components.

Stainless Steel Production in Zimbabwe

Zimbabwe is endowed with the key elements required in the production of stainless steel, namely, chromium (as ferrochrome) and nickel (as ferronickel).

Chromium is one of the most versatile and widely used alloying elements in steel. Chromium is an irreplaceable constituent in all stainless steels, with over 70 per cent of all chromium used in steelmaking being consumed in the various stainless steel grades (AMG Vanadium, 2009). It imparts corrosion and oxidation resistance, is a mild hardenability agent, improves wear resistance and promotes the retention of useful strength levels at elevated temperatures (AMG Vanadium, 2009). Other applications of chromium include structural and construction alloy steels, tool steels, super-alloys and other specialty metals.

Zimbabwe currently exports chrome products as partially beneficiated ferrochrome alloys. Ferrochrome is a corrosion-resistant alloy of chromium and iron, with the chromium content ranging between 50 per cent and 60 per cent (MMCZ, undated). Three broad ferrochrome product exist, namely high carbon ferrochrome (HCFeCr), low carbon ferrochrome (LCFeCr) and ferro-silicon chrome (FeSiCr), with the carbon level being the most important parameter in determining the price differential between the various ferrochrome grades (AMG Vanadium, 2009). The high carbon ferrochrome (HCFeCr) is the most widely used chromium addition for the production of stainless and alloy steels (AMG Vanadium, 2009; MMCZ, undated).

Zimbabwe's known chromium resources are in the range of 1000Mt, containing on average 500Mt of chromium. Zimbabwe ferrochrome production has been traditionally dominated by four major smelters, namely ZIMASCO in Kwekwe, Maranatha, Zimbabwe Alloys in Gweru (currently under care and maintenance), Riochrome in Kadoma, and Oliken in Kwekwe. Additional production is now coming from new but fragmented small scale smelters which are exploiting the Great Dyke ores. Despite being ranked second in the world in terms of reserves, Zimbabwe currently has the capacity to produce only about 6 per cent of the global production (Parliament of Zimbabwe, 2013).

Nickel: Nickel is an indispensable alloying element in the production of stainless and alloy steels due to its properties as a solid solution strengthener, a mild hardenability agent, and promoter of low temperature toughness in steels (AMG Vanadium, 2009). Nickel is also commonly used in the production of super alloys and heat and corrosion resistant materials (AMG Vanadium, 2009). Because of its properties in improving alloys' resistance to corrosion, and the ability to withstand extreme temperatures, nickel bearing alloys are often used in severe conditions such as those found in chemical plants, petroleum refineries, jet engines, power generation, and offshore installations (USGS, 2012).

Zimbabwe's nickel primary production is dominated by Empress Nickel Refinery (ENR), Bindura Nickel Corporation (BNC), and is also produced as a by-product from the smelting and refining of platinum group metals (PGMs). Despite very little information being available on the life cycle behaviour and the recycling capability of nickel and nickel products circulating in the economy, Zimbabwe is endowed with good nickel resources to sustain the long term production of stainless steel for both the local and regional markets (Parker, 2014).

Characteristics of the Zimbabwean Iron and Steel Industry

The iron and steel sector in Zimbabwe is characterized by a plethora of challenges. Most of these problems can be attributed to the poor technical, managerial and organizational capabilities which limit the ability of the local companies to compete in the global landscape.

The iron and steel sector in Zimbabwe is characterized by incomplete value chain participation and the poor value-in use of iron and steel intermediate products. This limits the ability and competitiveness of the downstream manufacturing companies to manufacture value-added products. Developing an innovation driven beneficiation framework to maximize the economic benefits within the iron and steel sector requires a holistic sector diagnosis by taking into account competitiveness of all the activities along the value chain, such as the primary smelting and refining, secondary processing, fabrication of intermediate products, machine and equipment building, assembly, and the production of end-market products. As a result, it is imperative to evaluate the technological, innovation and organizational capabilities in the Zimbabwean iron and steel sector, with the ultimate objective of building a sustainable and competitive sector.

Justification of the Study

The ZIM ASSET policy thrust on resuscitating the domestic iron and steel sector demonstrates the need to develop holistic technological and innovation capabilities driven beneficiation framework for the iron and steel sector (Government of Zimbabwe, 2014). The performance of the iron and steel sector in Zimbabwe has catalytic effects on the downstream industries' capability to generate value-added products. Huh (2010) established a causal relationship between the total steel consumption and GDP, with the two parameters having a long term bi-directional relationship in terms of the flat products consumption and the manufacturing gross domestic product (Huh, 2010). As a result, developing a conceptual framework to assess the current level of technological and innovative capabilities is the entry step towards building an innovation-driven iron and steel beneficiation roadmap.

Understanding the technological and innovation capabilities is crucial for the companies to identify the key potential growth areas, and institute proactive and/ or remedial measures to improve on their competitive position in terms of products and costs (Muller, 2005; Gerybadze, 2010; Saunila, 2013; OECD, 2005). Therefore, the focus of current study is to assist the respective organizations identifying the industry and firm-level strategic imperatives towards managing and developing sustainable technological and innovative solutions. The study also cascades into the development of national and regional mineral beneficiation policies, and is beneficial in addressing the exogenous and intrinsic factors affecting the competitiveness of the iron and steel sector within the regional and global benchmarks (Government of Zimbabwe, 2014; UNECA, 2013). The study is therefore important for developing such indicators for benchmarking the industry performance, and in promulgating the policies and measures that foster innovation and technology resilience.

Zimbabwe, like most developing countries, lacks a holistic framework for evaluating sector specific technological and innovation capabilities. The most common approach adopted globally for evaluating technical and innovation capabilities is through national innovation surveys (OECD, 2005). It is commonly agreed that the results from national and business sector innovation surveys provide a rich framework for policy implementation and intervention measures across the whole economy (Human Resources Research Council, 2008; OECD, 2005). Therefore, this study does not only contribute towards building the body of knowledge for evaluating the iron and steel sector specific technological and innovation capabilities within the local context, but also plays a critical role in identifying the key technological and innovation capability gaps within the broader iron and steel sector. Finally, the findings from this study also cascades into developing a conceptual framework for crafting a holistic innovation driven minerals beneficiation framework.

Objectives of the Study

The broad objective was to highlight the critical technical, innovation, managerial and organizational capability gaps required to develop a capabilities-driven beneficiation framework for the iron and steel industry in Zimbabwe. The secondary objective is to analyze the technological and innovation trends and drivers, the capabilities of measurement metrics, and the critical success factors to the current technological and innovation management strategies within the iron and steel sector in Zimbabwe. Finally, recommendations were made to integrate the sector specific technological and innovation best practices into the broader national beneficiation framework.

Methodology

This study adopted a qualitative based case study survey to analyze the current technological and innovation capabilities, and the key technological and innovation trends and drivers within the iron and steel industry in Zimbabwe. The instruments were developed as a hybrid of the OECD-CIS harmonized methodology, the Boston Consulting Group, and the Price Waterhouse Coopers methodologies for innovation surveys (OECD, 2005; EuroStat, 2013; Wagner, 2013; Wortler, 2010; Price Waterhouse Coopers, 2013).

Questionnaires soliciting for the general information and demographics, technological capabilities, innovation activities, innovation capabilities, technological and innovation capability measurement metrics, and the creativity and skills were distributed to the respondents. Furthermore, questions evaluating the level of beneficiation and value addition of the iron and steel products and raw materials were developed. Key informants interviews were applied to augment data collection from policy making and regulatory authorities, academia, and other related support institutions.

The study focused specifically on the key companies in the iron and steel sector in Zimbabwe. The companies were selected based on the level of activity and the criticality within the iron and steel sector. A total of sixty questionnaires were distributed to senior management (N=10), middle management (N=15), senior technical operatives (N=15), middle level technical operatives (N=15), and senior non-technical operatives (N=10). Additional on-line questionnaires were distributed to technical respondents within the region. The performance of the local companies was benchmarked against their regional counterparts based on an online questionnaire distributed to technical professionals in the region.

Data analysis was carried out based on the analytical framework adopted from the global surveys cited in literature (OECD, 2005; EuroStat, 2013; Wagner, 2013; Wortler, 2010; Price Waterhouse Coopers, 2013). Figure 2.2 shows analytical framework adopted in the study.

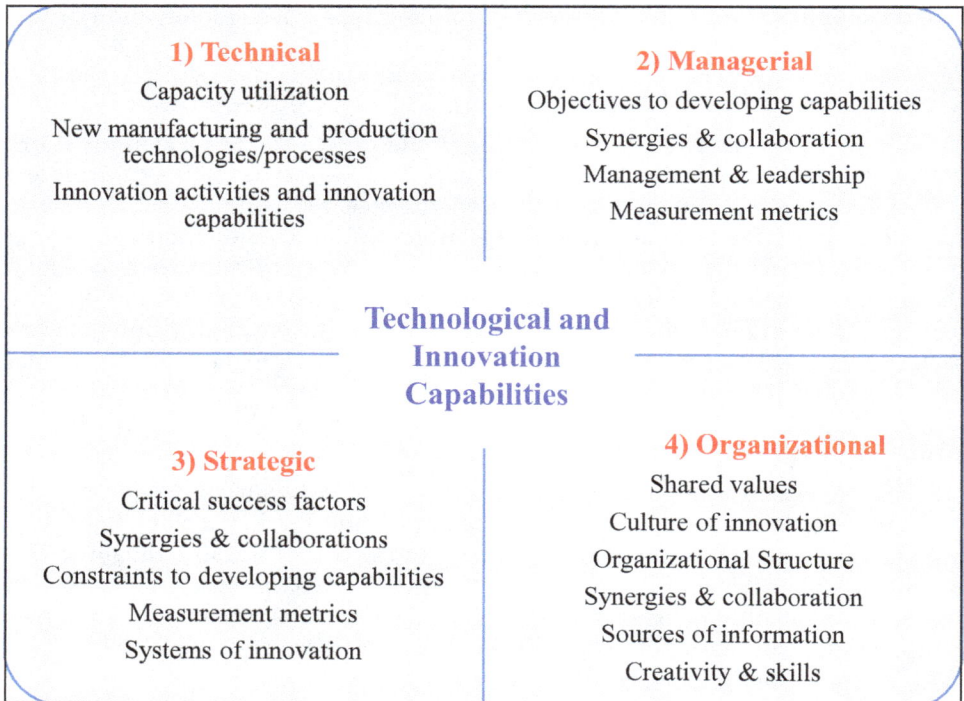

1) Technical
Capacity utilization
New manufacturing and production technologies/processes
Innovation activities and innovation capabilities

2) Managerial
Objectives to developing capabilities
Synergies & collaboration
Management & leadership
Measurement metrics

Technological and Innovation Capabilities

3) Strategic
Critical success factors
Synergies & collaborations
Constraints to developing capabilities
Measurement metrics
Systems of innovation

4) Organizational
Shared values
Culture of innovation
Organizational Structure
Synergies & collaboration
Sources of information
Creativity & skills

Figure 2.2: Analytical Framework for Measuring the Technical, Managerial, Organizational and Strategic Capabilities Attributes

Results and Discussion

Capacity Utilization and Industry Characteristics

The rate of capacity utilization is one of the key variables driving the cost competitiveness of the iron and steel industry in Zimbabwe. Figure 2.3 shows the capacity utilization for the industry. In general, the capacity utilization was generally below 50 per cent of the design and installed capacity for the period 2010 to 2013.

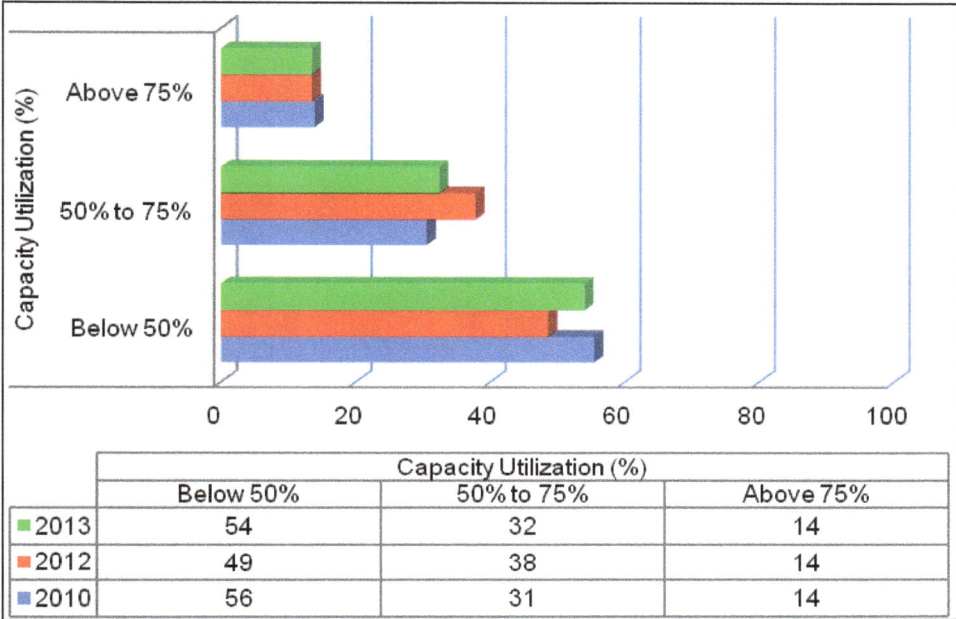

	Capacity Utilization (%)		
	Below 50%	50% to 75%	Above 75%
2013	54	32	14
2012	49	38	14
2010	56	31	14

Figure 2.3: Capacity Utilization as a Function of Design Capacity

Capacity utilization and the cost of production are usually inversely correlated, because the underutilization of capacity increases the cost of production through the absorption of budgeted costs (Hosen, 2011). The observed poor competitiveness of the general manufacturing sector in Zimbabwe is mostly attributed to the low levels of capacity utilization (Confederation of Zimbabwean Industries, 2013). With increased competition from cheap imports of iron and steel products, the local producers of iron and steel products are left with limited strategic choices but to compete on cost competiveness in order to drive profitability utilization (Confederation of Zimbabwean Industries, 2013).

The factors affecting the observed capacity utilization for the companies within the iron and steel sector were evaluated. Table 2.1 shows the impact rating for the factors affecting the observed capacity utilization for the companies within the iron and steel sector. The major constraints include obsolete plant and equipment, high operating costs, working capital constraints, poor demand and poor markets, and reduced capacity due to downsizing. The iron and steel sector is characterized by obsolete equipment and high operating costs as a result of low reinvestment and

Table 2.1: Impact Rating of the Factors Affecting the Observed Capacity Utilization

Reason	Critical to Most Critical	Somewhat	Not Critical	Responded
Increased demand and good market conditions	27 per cent (n=8)	30 per cent (n=9)	43 per cent (n=13)	30
Poor demand and bad market conditions	73 per cent (n=22)	17 per cent (n=5)	10 per cent (n=3)	30
Improved working capital availability	15 per cent (n=5)	35 per cent (n=12)	50 per cent (n=17)	34
Working capital constraints	77 per cent (n=27)	14 per cent (n=5)	9 per cent (n=3)	35
New and improved plant and equipment	20 per cent (n=6)	23 per cent (n=7)	57 per cent (n=17)	30
Obsolete plant and equipment	85 per cent (n=29)	12 per cent (n=4)	3 per cent (n=1)	34
Increased capacity utilization due to expansion	19 per cent (n=6)	26 per cent (n=8)	55 per cent (n=17)	31
Reduced capacity utilization due to downsizing	69 per cent (n=25)	17 per cent (n=6)	14 per cent (n=5)	36
High operating costs	80 per cent (n=28)	14 per cent (n=5)	6 per cent (n=2)	35
Favorable operating costs	13 per cent (n=4)	34 per cent (n=11)	53 per cent (n=17)	32

capacity building over the years prior to, and after the hyperinflationary period. The constraints to capacity utilization such as poor demand and poor markets for beneficiated products indicate the inherent structural weaknesses within the broader manufacturing, and these include low levels of export capacity and access to regional and international markets.

However, in some cases, the capacity utilization had increased in the period in question due to favourable operating costs, increased capacity utilization as a result of expansion, new and improved plant and equipment acquisition, improved working capital availability and increased demand and good market conditions.

Industry attributes such as capacity utilization, technology competitiveness, process efficiency, and the quality of products and raw materials provide an indicative benchmark for measuring the industry performance and profitability. As a result, the industry attributes were evaluated benchmarked on competitiveness in Zimbabwe (ZW) and to regional competitors (REG). Table 2.2 shows the attributes rating of the local iron and steel companies when benchmarked against the local and regional counterparts.

Although there was confidence that the quality of raw materials and products were good to very good relative to their local competitors, the peer to peer ranking in terms of attributes such as technology competitiveness, capacity utilization and process efficiency was generally poor. Generally, the local companies rated poorly in all the attributes evaluated when benchmarked against their regional (REG) counterparts. Common reasons cited for the poor competitiveness in terms of technology competitiveness, capacity utilization and process efficiency included the low level of new investment in capital equipment to modernize or expand manufacturing operations, the low country investment competitiveness due to low investor confidence in the broader economic environment, the liquidity challenges currently affecting the local financial sector, and the virtual collapse of the iron and steel sector during the hyperinflationary period.

The rating of technological capabilities was assessed with respect to local competitors (Zimbabwe), regional, and global competitors. The technological capabilities were marginally competitive to not competitive at all with respect to regional and global competitors, respectively. Figure 2.4 shows the technological capabilities rating benchmarked against the regional and global competitors.

New Manufacturing and Production Technologies and Methods

The most innovative companies continuously reinvest in building long term technical capabilities (Gerybadze, 2010; Wagner, 2013; Saunila, 2013). Despite the technological capabilities being predicated by the rate of incorporation of new and improved manufacturing and technologies, only a marginal number of local iron and steel companies had acquired new technological capabilities enhancing interventions such as new manufacturing, production techniques, new plant and machinery, and new manufacturing and production processes. Figure 2.5 shows the percentage of the annual turnover that was reinvested in capacity building and competitiveness enhancing initiatives in the period 2010 to 2013.

Table 2.2: Attributes Rating Benchmarked against the Local and Regional Counterparts

Industry Attribute	High to Very High	Marginal	Poor	Total
Product quality (ZW)	77 per cent (n=23)	13 per cent (n=4)	10 per cent (n=3)	30
Raw material quality (ZW)	80 per cent (n=24)	13 per cent (n=4)	7 per cent (n=2)	30
Process efficiency (ZW)	17 per cent (n=5)	57 per cent (n=17)	27 per cent (n=8)	30
Capacity utilization (ZW)	17 per cent (n=5)	57 per cent (n=17)	27 per cent (n=8)	30
Technology competitiveness (ZW)	17 per cent (n=5)	53 per cent (n=16)	30 per cent (n=9)	30
Product quality (REG)	13 per cent (n=4)	30 per cent (n=9)	57 per cent (n=17)	30
Raw material quality (REG)	20 per cent (n=6)	50 per cent (n=15)	30 per cent (n=9)	30
Process efficiency (REG)	10 per cent (n=3)	30 per cent (n=9)	60 per cent (n=18)	30
Capacity utilization (REG)	7 per cent (n=2)	37 per cent (n=11)	57 per cent (n=17)	30
Technology competitiveness (REG)	10 per cent (n=3)	30 per cent (n=9)	60 per cent (n=18)	30

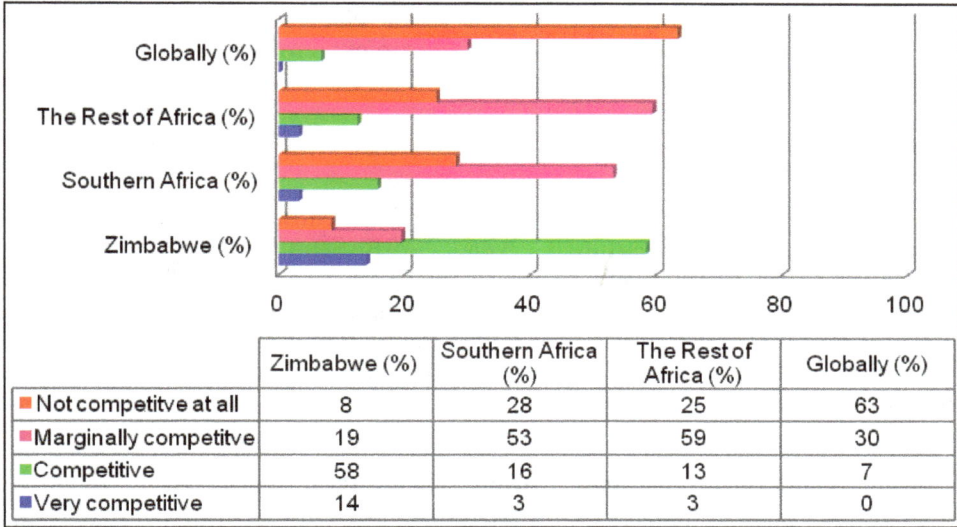

	Zimbabwe (%)	Southern Africa (%)	The Rest of Africa (%)	Globally (%)
■ Not competitve at all	8	28	25	63
■ Marginally competitve	19	53	59	30
■ Competitive	58	16	13	7
■ Very competitve	14	3	3	0

Figure 2.4: Technological Competitiveness Rating

In general, most of the companies had reinvested less than 10 per cent of their annual turnover on competitiveness building strategies such as new plant and machinery, human capital and skills development, process improvement initiatives, and on other modernizations such as demand side energy management

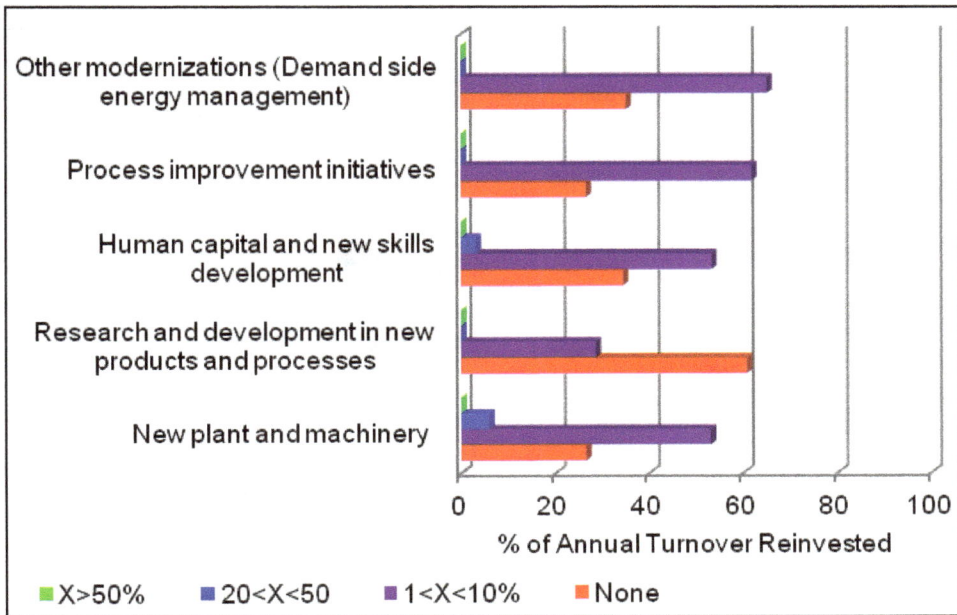

Figure 2.5: Percentage of Annual Turnover Reinvested in Capacity and Competitiveness Building

initiatives. The observed trend in the amount of retained earnings reinvested in capacity building clearly indicates that the companies were focusing on quick gains technological capabilities building strategies. Because the iron and steel making processes are energy intensive, the demand side energy management becomes a critical parameter in the final product cost model (Confederation of Zimbabwean Industries, 2013).

The observed trend of reinvestments in R&D is in contrast to the common trends identified by scholars in literature (Gerybadze, 2010; Saunila, 2013). Organizations that continuously invest in R&D to expand their technological capabilities base often attain stable growth and strong financial performance (Gerybadze, 2010; Saunila, 2013). According to the cycle of innovation, the high margins and above-average returns attained by innovative organizations are ploughed back to be reinvested in more R&D activities than competitors, thereby such organizations to manage new product development pipeline in perpetuity (Gerybadze, 2010).

Strategic Considerations given Unlimited Financial Resources

From the foregoing observations, financial resources (*i.e.* capital and working capital) constraints were cited as one of the major constraints to achieving the full capacity utilization in the local sector. Therefore, the respondents were tasked to rank their strategic priorities given unlimited financial resources. Figure 2.6 shows the top most priorities given unlimited financial resources.

The respondents prioritized new production methods, process development and redesign, continuous improvement of existing processes and technologies, and

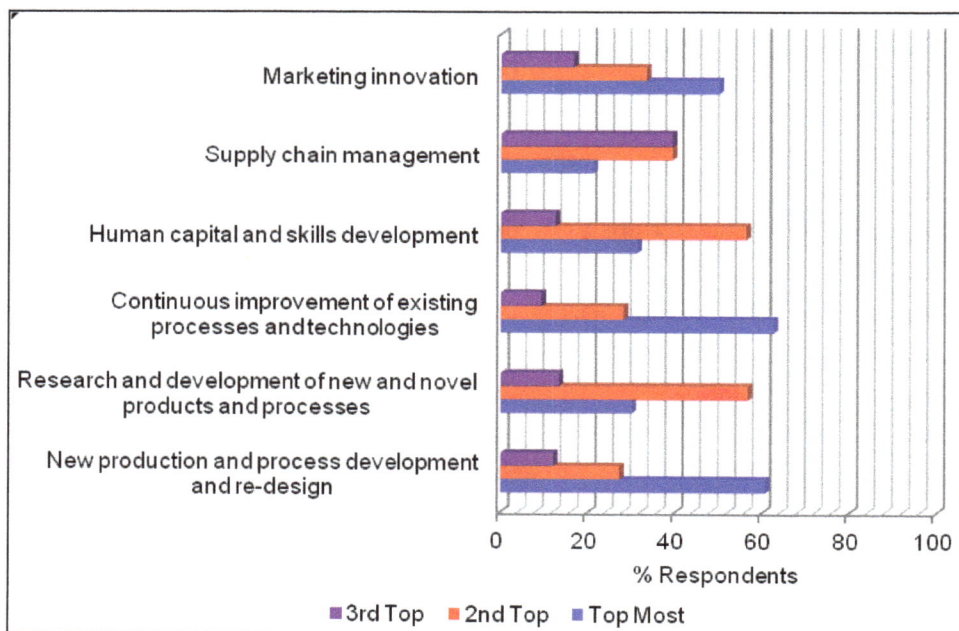

Figure 2.6: Top most Priorities given Unlimited Financial Resources

marketing innovation. In contrast, long term capacity building initiatives such as research and development of novel products and processes, and human capital and skills development were given second priority. The results indicate a deliberate strategic thrust towards building short term competitiveness within the sector.

Critical Success Factors in Planning and Execution of Capabilities Enhancing Activities

The most innovative companies in the world strongly leverage on the critical success factors such as the ability to finance the reduction to practice of innovations, building a culture of innovation, collaboration and attracting and retaining talent to drive technological and innovation capabilities (*BCG, 2010; KPMG, 2012; BCG, 2013*). As a result, the local companies were ranked on their ability to overcome some of these critical success factors in the planning and execution of technological and innovation capabilities enhancing activities.

As shown in Figure 2.7, the majority of the companies faced challenges in financing the reduction to practice and commercialization of innovative ideas, building a culture of innovation within the organization, attracting and retaining skills, finding the right collaborative partners, and in the rapid deployment of innovative ideas at the right scale. As a result, the companies were failing to leverage on the common characteristics of innovative organizations through developing dynamic technological and innovation capabilities. For example, most companies rated poorly in growth catalysts such as leveraging on strong intellectual property, organizational growth and strategic flexibility, developing a culture of continuous learning and organizational renewal, developing a strong culture of research and development.

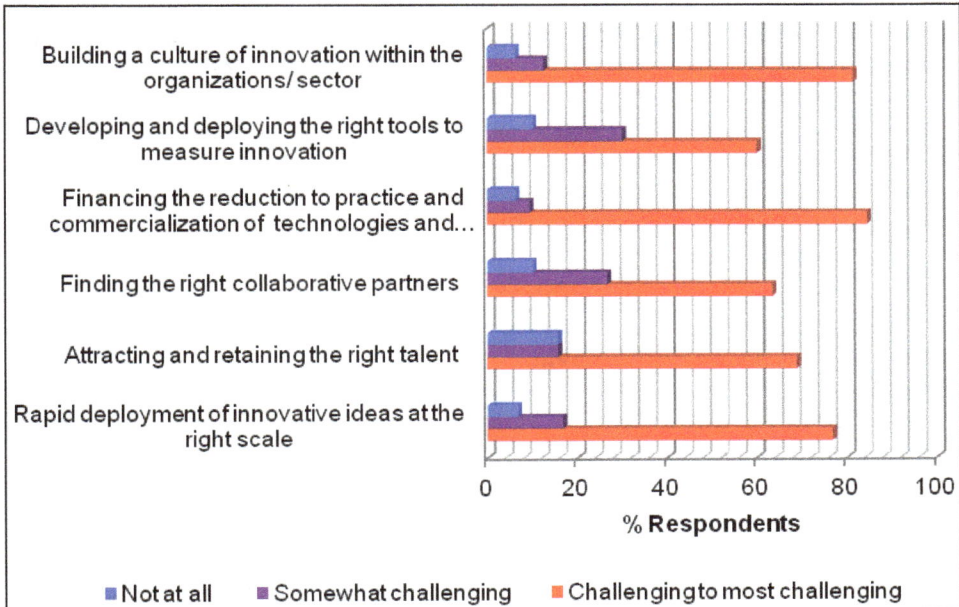

Figure 2.7: Rating of Critical Success Factors

Several studies have been conducted on how the most innovative companies in the world strongly leverage on these critical success factors in driving technological and innovation capabilities (Wortler, 2010; KPMG, 2012; Wagner, 2013). Yam *et al.* (2004) defined the manufacturing capability of a company as the ability to reduce to practice the results and efforts of R&D and innovation activities in developing products which meet the design criteria and satisfy both the present and future market needs (Yam, 2004). Yam *et al.* (2004). further emphasized the importance of linking the manufacturing capabilities and market intelligence, that is, the information gathered from the market intelligence continuously fed back into developing and manufacturing more robust and advanced products that meet or surpasses customer expectations and competitors' products (Yam, 2004). Based on the definition of innovative organizations, the local companies had challenges in sustaining the organizational capabilities through developing a strong culture of R&D and organizational learning and renewal as such organizational attributes affect the pace of innovation processes through technological innovation enabling structures and systems (Tseng, 2012; Yam, 2004; Wagner, 2013; Wortler, 2010).

Innovation Activities and Innovation Capabilities

In most cases, the technological and innovation capabilities are predicated on how the respective organizations manage and execute their innovation activities. Based on the CIS Harmonized Innovation Surveys (EuroStat, 2013), the companies were evaluated whether they had introduced any new processes, products, marketing, service and organizational innovations in the period between 2010 and 2013. Table 2.3 summarizes the type of innovations introduced.

The results show that only a marginal number (below 50 per cent) of companies had introduced new innovations in terms new or significantly improved processes, supporting activities, improved marketing or distribution methods. However, the companies fared fairly well in terms of the introduction of organizational innovations such as new methods of organizing work responsibilities and new methods for organizing external relations with other firms. The technological and innovation capabilities are a result of how the companies manage and execute their innovation activities, and the innovation savviness of an organization is a function of the rate of introduction of new innovations. In contrast to the observed trend, the most innovative companies are characterized by rapid deployment of innovative ideas at the right scale and speed (Wagner, 2013).

The innovations shown in Table 2.4 were further categorized into incremental, radical and breakthrough innovations. Additional information on how the innovations were developed, and the novelty of these innovations to the market (*i.e.* local or peripheral to the company), in Zimbabwe or in the world, was solicited. Table 4 shows the classification of the new innovations introduced.

From the results, most of the process, product and marketing innovations were incremental innovations, while service and organizational innovations distributed among incremental, radical and breakthrough innovations. A significant portion of the process innovations were either developed by the companies in collaboration with other institutions and were categorized as either modified or

Table 2.3: Summary of the Types of Innovations Introduced

Type of Innovation	YES	NO	Respondents
Process innovations	New or significantly improved processes PROC1	41 per cent (n=12)	59 per cent (n=17)	29
	New or significantly improved supporting activities e.g. maintenance systems computing –PROC2	39 per cent (n=11)	61 per cent (n=17)	28
Marketing	New or significantly improved marketing or distribution methods -MKTNG	37 per cent (n=11)	63 per cent (n=19)	30
Product	New or significantly improved goods -PROD	43 per cent (n=13)	57 per cent (n=17)	30
Service	New or significantly improved services -SER	42 per cent (n=13)	58 per cent (n=18)	31
Organizational innovations	New business practices for organizing procedures –ORG1	38 per cent (n=11)	62 per cent (n=18)	29
	New methods of organizing work responsibilities and decision making-ORG2	57 per cent (n=17)	43 per cent (n=13)	30
	New methods for organizing external relations with other firms or institutions-ORG3	57 per cent (n=17)	43 per cent (n=13)	30

Table 2.4: Classification of the New Innovations Introduced

Type	Classification				Who Developed them				
	Incremental	Radical	B/through	Total	Own	Collaborated	Modified	Adopted	New in
Process	12 (100%)	0	0	12	2 (18%)	6 (55%)	2 (18%)	1 (9%)	
Product	13 (100%)	0	0	13	0	2 (15%)	11 (85%)	0	ZW: n=3 (23%)
Marketing	11 (100%)	0	0	11	7 (64%)	3 (27%)	1 (9%)	0	
Service	9 (53%)	5 (29%)	3 (18%)	17	7 (41%)	3 (18%)	6 (35%)	1 (6%)	MKT: n=6 (35%)
Organization	9 (52%)	5 (24%)	3 (24%)	17	9 (53%)	5 (29%)	0	3 (18%)	

adopted innovations while most product innovations were a modification of the products originally developed by other institutions. Marketing innovations were predominantly from the effort of the innovating companies themselves, while organizational innovations were as a result of own efforts, collaboration, and adopting existing innovations. In terms on the novelty of the product innovations, only a marginal number of the companies had product innovations being new to Zimbabwe, and service innovations being novel to the market.

The observed technological and innovation capabilities in terms of process, product and marketing innovations were mostly incremental innovations such as the incremental changes to existing attributes, improvements or extensions. Incremental innovations are a result of continuous improvement of existing processes and products, and are mostly focused on costs or feature improvements in product, services, marketing or business models, and are characterized by low risk and uncertainty during their deployment (Australian Institute for Commercialization, 2009; Tushman, 1996).

Overwhelming evidence in literature support that notion that it is the rate of introduction of new innovations, especially in terms of radical and breakthrough innovations, that often add significant new value to the marketplace and serve as the foundation for subsequent technological developments (Schumpeter, 1934; Ahuja, 2001; Slocum, 2008). In addition, literature supports that radical and breakthrough innovations are usually a result of conceited efforts through research and development, serendipity, ingenuity and ambidexterity of innovating individuals or organizations, they are also associated with high costs, risks and uncertainty (Tushman, 1996). However, none of the companies surveyed had introduced any radical or breakthrough process, product or marketing innovations.

Objectives and Strategies in Developing Technological Innovations

It is important to align the companies' long term growth objectives to technological and innovation capabilities. As such, the importance rating of the objectives towards developing technological innovations was evaluated. The results in Figure 2.8 show that the top most important objectives in developing technological innovations were improving the flexibility of production processes, improving the quality of products, and increased productivity. Other important objectives included entering new markets and replacing outdated technologies. Improving the environmental footprint was also prioritized due to the stricter monitoring and regulation by environmental watchdogs, increased environmental awareness by the company employees, and the need to be a good corporate citizen.

Most Smportant strategies in driving innovation capabilities

Different companies employ different strategies to managing and driving their innovation capabilities. Table 2.5 shows the most effective strategies in driving the technological and innovation capabilities, and these include driving these innovation activities through the use of dedicated innovation and R&D teams and multifunctional teams. Top management support was also crucial in driving innovation in most companies.

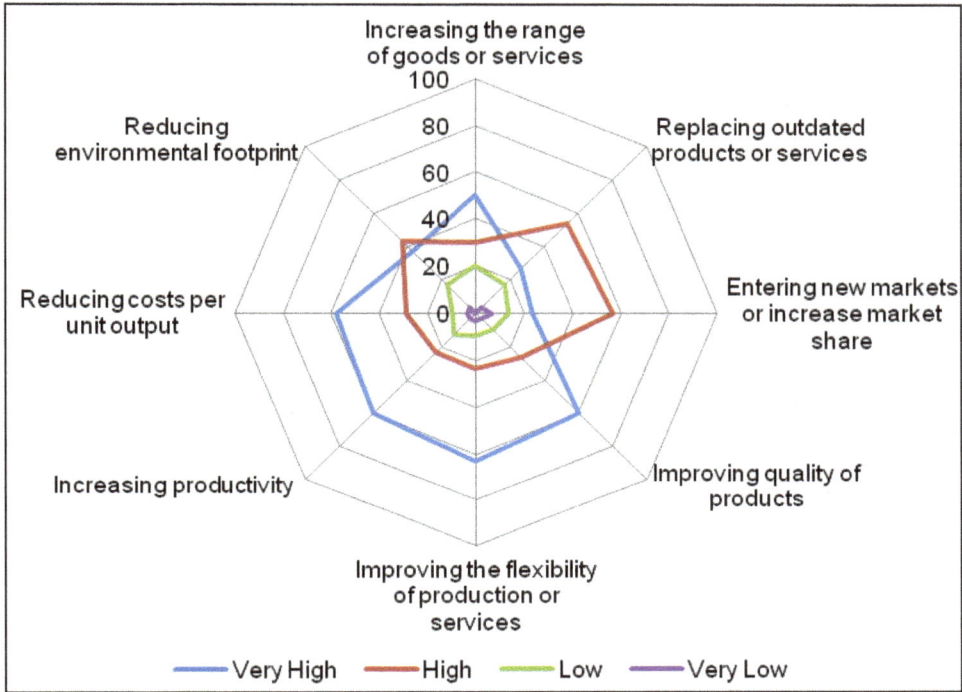

Figure 2.8: Importance Rating of the Objectives Towards Developing Technological Innovations

Organizational systems play an important role in building and fostering technological and innovation capabilities, and such systems are predicated on the capacity to constitute well-established organizational structures, cultivate innovation-centric organizational culture and vision, and impacting on the pace of innovation processes through technological innovation enabling structures and systems (Tseng, 2012; Yam, 2004). Innovative organizations create a cross-functional mechanisms and diversity to manage innovation activities from ideation to implementation and launch of new products through the use of dedicated and cross-functional teams to drive innovations (Cooper, 2007).

The respondents (81 per cent) concurred that top management support is crucial in driving innovation in organizations. Management support is important in building and fostering dynamic technological and innovation capabilities through a complete alignment of the innovation objectives to the company wide strategies (Wagner, 2013; Gerybadze, 2010). According to Wagner *et al.*, 2013, strong innovators have top management that is committed to innovation as a competitive advantage, and managed a portfolio of innovative initiatives, and leveraged on strong customer focus, and insist on strong processes as a driver of strong performance (Wagner, 2013). When top management commitment is complimented with other factors such as strategic direction, organizational capabilities and dynamic interaction

Table 2.5: Most Effective Strategies in Driving Technological and Innovation Capabilities

	Effective	Somewhat Effective	Not at all	Total N	Ranking
Discrete product areas and business units drive their own innovations	10 (29 per cent)	20 (57 per cent)	5 (14 per cent)	35	4
Through formal innovation structures in the discrete business units	13 (36 per cent)	21 (58 per cent)	2 (6 per cent)	36	5
Innovation is driven across the whole organization through the use of multifunctional teams	30 (83 per cent)	4 (11 per cent)	2 (6 per cent)	36	2
Top management support is crucial in driving innovation	31 (81 per cent)	3 (8 per cent)	2 (6 per cent)	36	3
Innovation is driven across the whole organization through the use of dedicated innovation and R&D teams	29 (83 per cent)	5 (14 per cent)	1 (3 per cent)	35	1

with the firms' ecosystem, innovating organizations are therefore able to build dynamic capabilities based on strong managerial capabilities at both the corporate as well as the business unit level (Gerybadze, 2010). In addition, the most innovative businesses made innovation a priority at board level, making capital available to support innovative ideas (PwC, 2013).

Constraints to Developing Technological Innovations

In collorary to the respondents being aware of the criticality of building innovation capabilities, there was no observed correlation between the respondents' knowledge of the criticality of building capabilities with the companies' current technological and innovation capabilities. As a result, the constraints to building technological and innovation capabilities were evaluated and categorized as cost factors, knowledge, market conditions, organizational culture, risk factors, and other reasons such as preference to adopt open market innovations. Table 2.6 shows the importance rating of the constraints to embarking on innovation activities.

From Table 2.6, the most binding constraints to innovation activities were difficulty in finding cooperating partners, lack of internal and external sources of finance to embark on innovation activities, high innovation costs, uncertain demand for innovation products, high risks associated with new innovations, and the companies' preference to adopt technology on the open market. Such severity of the cited binding constraints was synonymous with the business operating environment during the period under review.

Sources of Information

Leveraging on the sources of information for innovation activities is critical for innovating organizations. As a result, the importance of sources of information was evaluated based on the company itself (internal), information from the market, institutional sources, and trade associations. Table 2.7 summarizes the findings from the survey.

The companies in Zimbabwe are mostly utilizing private consultants and commercial research and development laboratories as important sources of information to innovation activities. Other important sources of information include suppliers, and customers and clients. Public universities and public R&D institutions were moderate sources of information. In contrast to the most innovative organizations that leveraged on scientific journals and technology repositories and hatcheries as precursors to new technology development and technology transfer, the scientific journals and publications, technology repositories and hatcheries were the least utilized sources of information by the companies within the Zimbabwean iron and steel.

Collaboration and synergies among the sources of information and the companies' innovation teams is important for the dissemination of information pertaining to innovations. It is commonly accepted that the linkages the organization has within its eco-system through open innovation systems, and the capability to use external knowledge networks to drive radical innovation, influences the innovative capability with respect to competitors (McAdam, 2014). Table 2.8 shows

Table 2.6: Importance and Rating of the Constraints to Embarking on Innovation Activities

Constraints to Innovation Activities	Degree of Importance			Total	Rank
	High	Medium	Low		
Cost					
Lack of funds within your organization (CST1)	30 (85 per cent)	5 (14 per cent)	1 (3 per cent)	36	
Lack of finance from external sources (CST2)	32 (80 per cent)	7 (18 per cent)	1 (3 per cent)	40	
Innovation costs too high (CST3)	31 (80 per cent)	6 (15 per cent)	2 (5 per cent)	40	
Knowledge					
Lack of qualified personnel (KNO1)	18 (43 per cent)	22 (52 per cent)	2 (5 per cent)	42	
Lack of information on technology (KNO2)	12 (29 per cent)	23 (55 per cent)	7 (17 per cent)	42	
Difficulty in finding cooperating partners (KNO3)	34 (85 per cent)	5 (13 per cent)	1 (3 per cent)	40	
Lack of information on markets (KNO4)	13 (33 per cent)	22 (55 per cent)	5 (13 per cent)	40	
Market					
Market dominated by competitors (MKT1)	30 (79 per cent)	5 (13 per cent)	3 (8 per cent)	38	
Uncertain demand for innovative products (MKT2)	32 (80 per cent)	5 (13 per cent)	3 (7 per cent)	40	
No market demand to justify the investment (MKT3)	26 (62 per cent)	11 (26 per cent)	5 (12 per cent)	42	
Culture					
Organizational culture stifles innovation (ORG1)	30 (75 per cent)	7 (18 per cent)	3 (7 per cent)	40	
No rewarding system for innovations (ORG2)	31 (78 per cent)	7 (18 per cent)	2 (4 per cent)	40	
Risk					
New innovations are too risky (RISK)	32 (80 per cent)	5 (13 per cent)	3 (7 per cent)	40	
Other Reasons					
No need due to prior innovations (OTH1)	6 (15 per cent)	12 (31 per cent)	21 (54 per cent)	39	
Prefer to adopt technology on the market (OTH2)	31 (80 per cent)	5 (13 per cent)	3 (7 per cent)	40	

Table 2.7: Importance of Information Sources to Innovation Activities

Importance of Information Sources	Degree of Importance			Total	Rank
	High	Medium	Low		
Internal Within company (INT)	12 (46 per cent)	13 (50 per cent)	1 (4 per cent)	26	Medium
Market Suppliers of equipment, materials, (MKT)	22 (76 per cent)	6 (21 per cent)	1 (3 per cent)	29	High
Competitors (COMP)	9 (29 per cent)	17 (55 per cent)	5 (16 per cent)	31	Medium
Customers and clients (CUST)	23 (74 per cent)	7 (23 per cent)	1 (3 per cent)	31	High
Institutional Consultants, private R&D and commercial labs (PRIVTCONSUL)	28 (78 per cent)	7 (19 per cent)	1 (3 per cent)	36	High
Universities and colleges (VARS)	3 (11 per cent)	15 (56 per cent)	9 (33 per cent)	27	Medium
Public R&D institutes (PUBCONSUL)	3 (10 per cent)	15 (52 per cent)	11 (38 per cent)	29	Medium
Technology repositories and hatcheries (TECHREPOS)	3 (11 per cent)	9 (32 per cent)	16 (57 per cent)	28	Low
Trade associations Conferences, trade fairs, (CONF)	8 (28 per cent)	16 (55 per cent)	5 (17 per cent)	29	Medium
Scientific journals and publications (JOUR)	3 (10 per cent)	15 (48 per cent)	13 (42 per cent)	31	Low
Professional and industry associations (PROFAS)	12 (39 per cent)	16 (52 per cent)	3 (10 per cent)	31	Medium

Table 2.8: Effectiveness of Synergies and Collaboration with Sources of Information

Effectiveness of Synergies and Collaboration	Most Effective	Effective	Somewhat Effective	Not at all	Total	Ranking
Universities and colleges	0 (0 per cent)	5 (16 per cent)	17 (55 per cent)	9 (29 per cent)	31	5
Public R&D institutes	0 (0 per cent)	2 (8 per cent)	16 (64 per cent)	7 (28 per cent)	25	6
Consultants, private labs or private R&D Institutions	9 (27 per cent)	17 (52 per cent)	5 (15 per cent)	2 (6 per cent)	33	2
Suppliers of equipment, materials, components, etc	11 (31 per cent)	17 (47 per cent)	7 (19 per cent)	1 (3 per cent)	36	3
Clients or customers	9 (30 per cent)	16 (53 per cent)	5 (17 per cent)	0 (0 per cent)	30	1
Professional and industry associations	1 (4 per cent)	13 (50 per cent)	11 (42 per cent)	1 (4 per cent)	26	4

the effectiveness of the synergies and collaborations with sources of information for innovation activities.

The collaboration between companies and public institutions was somewhat effective. As is the case in most developing economies, the flow of information pertaining to innovation within national innovation systems is fragmented, and is often characterized by weak or complete absence of linkages between public scientific institutions and manufacturing enterprises (OECD, 2005). These links between the manufacturing companies and the science and technology institutes in developing countries are weak due to poor working relationships and understanding between manufacturing companies and these institutions, and also due to the misplaced perceptions on the public research institutes' abilities to carry out any meaningful industrial research (Lall, 2004; Pietrobelli, 2006). As a result, the weak or absent linkages challenge the capacities of iron and steel companies to overcome technological related problems and/or to drive innovations leading the companies to rely mostly on the acquisition of embodied technology (OECD, 2005; UNCTAD, 2007). This trend was confirmed through key informant interviews where the respondents expressed low confidence in the public research institutes' ability to solve manufacturing related problems.

Technology and Innovation Measurement Metrics

In an attempt to fully evaluate the dynamics of the current technological and innovation capabilities, the type and effectiveness of the technological and innovation capabilities measurement and benchmarking metrics were evaluated. Table 2.9 shows the type and effectiveness of the measurement metrics applied by the companies.

The financial metrics were the most commonly used technological and innovation capabilities measurement metrics by companies within the iron and steel in Zimbabwe. The use of balanced financial metrics such as the Balanced Scorecard and Return on Investment addressed the use of both financial and non-financial measures during corporate planning and performance measurement systems (Kaplan, 1996; Khomba, 2011). In addition, the balanced measurement metrics provide for the information that covers all relevant areas of corporate performance measurement systems (Kaplan, 1996; Khomba, 2011)

The companies were also using other performance measurement metrics such as the number of prototypes commercialized in a year, R&D expenditure, and market share compared to competitors. The use of generic measurement metrics such as project portfolio and number of active projects, and the number of ideas submitted by employees were the least preferred. Extensive literature supports the use of balanced performance measurement and management systems through the use of diverse financial and non-financial measures to enhance a balanced decision-making process for the management and stakeholders within the iron and steel (Neely, 2000; Neely, 2005).

Table 2.9: Type and Effectiveness of the Capabilities Measurement Metrics

	Effective	Somewhat Effective	Not at all	Total	Rank
R&D Expenditure	30 (83 per cent)	4 (11 per cent)	2 (6 per cent)	36	3
Patents and other IP	15 (38 per cent)	21 (53 per cent)	4 (10 per cent)	40	6
Financial metrics (Balanced Scorecard, ROI)	34 (85 per cent)	4 (10 per cent)	2 (5 per cent)	40	1
Number of citations and publications	7 (20 per cent)	9 (26 per cent)	19 (54 per cent)	35	8
Linkages, networks and other cooperation	11 (33 per cent)	17 (52 per cent)	5 (15 per cent)	33	7
Market share compared to major competitors	30 (81 per cent)	5 (14 per cent)	2 (5 per cent)	37	4
Project portfolio and number of active projects	5 (14 per cent)	18 (50 per cent)	13 (36 per cent)	36	10
Number of successful ideas submitted by employees	6 (16 per cent)	10 (27 per cent)	21 (57 per cent)	37	9
Percentage of sales from new products introduced in the past number of years	32 (78 per cent)	6 (15 per cent)	3 (7 per cent)	41	5
Number of prototypes commercialized per year	34 (85 per cent)	4 (10 per cent)	2 (5 per cent)	40	2

Creativity and Skills

Creativity and skills of employees determine the innovation capabilities in terms of idea generation and the reduction to practice of such innovative ideas. This section evaluated the suitability rating of the tertiary education curricula to skills requirements, and the most suitable methods of stimulating creativity among company employees. Table 2.10 shows the suitability rating of the tertiary education curricula to the iron and steel sector's needs.

Table 2.10: Suitability of the Tertiary Education Curricula to the Sector Needs

	Very Suitable	Suitable	Marginally Suitable	Not Suitable at all	Totals
Numeric Rating	5	15	15	1	36
Per cent Rating	14	42	42	3	100

Developing technological and innovation capabilities also requires the right skills and continuous development of human capital. Despite the poor synergies and collaboration between public academic institutions, the tertiary education curriculum was deemed suitable to very suitable to the skills needs of iron and steel sector. However, the main concerns raised were the lack of industry involvement in curricula development (industry concerns) and the poor collaboration between academia and industry. Propositions to constitute industry boards at academic, training, vocational and education institutions were made by some industry practitioners.

The most innovative organizations continuously engage and stimulate their employees to be creative through the right organizational systems and incentives (Wagner, 2013; Price Waterhouse Coopers, 2013). In addition, the degree of innovativeness of an organization is predicated on the creativity of employees within that organization. Table 2.11 evaluated the most suitable methods being employed to stimulate new ideas or creativity among the employees within the local iron and steel sector.

As shown in Table 2.11, job rotation among staff to different departments, multi-disciplinary and cross-functional teams, and financial incentives were the most suitable methods to stimulating creativity among staff. Other methods such as informal brainstorming sessions and non financial incentives were deemed less effective. The results show a balanced financial and non-financial approach in stimulating new ideas and creativity amongst employees. According to Price Waterhouse Coopers, using carefully designed reward and recognition systems to reinforce employee behaviour that encourages innovation drives innovation in successful organizations (PwC, 2013).

Beneficiation Capacity as a Function of Technological and Innovation Capabilities

The end result of studying the current technological and innovation capabilities is to evaluate the drivers and determinants of the value added processing of the iron

Table 2.11: Most Suitable Methods to Stimulate New Ideas or Creativity among Staff

	Most suitable	Suitable	Somewhat suitable	Not at all	Total	Ranking
Formal brainstorming sessions	3 (9 per cent)	12 (38 per cent)	16 (50 per cent)	1 (3 per cent)	32	4
Informal brainstorming sessions	2 (7 per cent)	11 (37 per cent)	15 (50 per cent)	2 (7 per cent)	30	5
Multi-disciplinary or cross-functional work teams	5 (21 per cent)	16 (53 per cent)	2 (7 per cent)	1 (3 per cent)	24	2
Job rotation of staff to different departments or other parts of the group	11 (37 per cent)	16 (53 per cent)	2 (7 per cent)	1 (3 per cent)	30	1
Financial incentives for employees to develop new ideas	18 (55 per cent)	9 (27 per cent)	5 (15 per cent)	1 (3 per cent)	33	3
Non financial incentives, *e.g.* public recognition	2 (7 per cent)	3 (10 per cent)	13 (43 per cent)	12 (40 per cent)	30	6

and steel products in Zimbabwe. As a result, the current the levels of beneficiation, value-added manufacturing, and or value in use of iron and steel raw materials and products was evaluated based on global benchmarks.

As shown in Table 2.12, only a marginal fraction of companies were producing fully value added products, with the majority producing either partially or non-value added products with respect to the regional and global standards. A cumulative 33 per cent of the respondents were either producing marginally value added products or exporting their products as raw or in un-beneficiated form.

Table 2.12: Degree of Value Addition Capacity of the Local Companies

Level	Respondents	Per cent Respondents
Fully value added final product	3	10
Producing partially value added intermediate products	17	57
Producing marginally value added products	6	20
Exporting as raw or un-beneficiated products	4	13
Column totals	30	100

As shown in Table 2.12, one of the pertinent problems affecting the local iron and steel sector is the lack of full value chain participation. While stainless steel is one of the most value added steel end products, there is no company that is producing stainless steel products in the country, implying that all the country's stainless requirements are being met by imports. As a result, exporting the intermediate iron and steel products while importing the more value added stainless steel and engineering products often results in an unsustainable trade deficit.

The binding constraints to the current level of value added processing of the iron and steel products were evaluated and were categorized as the lack of technical capacity to fully value add, lack of financial capacity to invest in value addition, market competition on value added products making further beneficiation unattractive, and lack of markets for fully value added products. The technical, financial and market constraints ranked as the most critical binding constraints to value addition and beneficiation within the sector. The results also show that the economic environment did not justify additional investment to further beneficiate the intermediate products, with the uncertainty over the socio-economic conditions prevailing in the country being the most critical.

Summary of Findings and Recommendations

Proactive strategies to maximize the utilization of available capacity using the available resources and capabilities were lacking in the iron and steel sector. Because capacity utilization and the cost of production are usually inversely correlated, the current underutilization of installed capacity increases the cost of production through the absorption of budgeted costs. (Hosen, 2011).

Because of the poor investment potential in Zimbabwe, the capital finance to build on technological capabilities is difficult to access. The low rating on the technological capabilities with respect to the regional and global competitors

Table 2.13: Binding Constraints to the Observed Level of Value Added Processing or Beneficiation

Reason	Important	Somewhat Important	Not Important	Total	Ranking
Lack of technical capacity to fully value add	26 (87 per cent)	3 (10 per cent)	1 (3 per cent)	30	1
Lack of financial capacity to invest in value addition	25 (86 per cent)	2 (7 per cent)	2 (7 per cent)	29	2
Lack of knowledge/skills to beneficiate further	16 (53 per cent)	11 (37 per cent)	3 (10 per cent)	30	6
There is no market for fully beneficiated products	19 (63 per cent)	8 (27 per cent)	3 (10 per cent)	30	5
Market competition makes further beneficiation unattractive	24 (83 per cent)	3 (10 per cent)	2 (7 per cent)	29	3
Exporting as raw or un-beneficiated makes more business sense	17 (57 per cent)	9 (30 per cent)	13 (7 per cent)	29	4
The economic environment does not justify additional investment to fully value add the products	18 (60 per cent)	11 (30 per cent)	3 (10 per cent)	28	5

was militating against the competitiveness of the local producers. It is therefore imperative that the companies continuously focus their efforts in continuous improvement initiatives of the existing processes in order to improve the current cost and technological competitiveness.

Because iron and steel making is energy intensive, high energy costs often contribute to the overall cost competitive pressures for the producers. As a result, a conceited investment effort in more energy efficient manufacturing systems and robust demand side energy management initiatives on existing manufacturing systems was required.

Government support is essential in growing the iron and steel sector. The Government support through financial and non-financial incentives such as subsidies on competitive capital equipment and technologies, tax incentives on new value addition and beneficiation investments, and in technical skills manpower development.

The existing management strategies were not sufficient in addressing the long term growth and competitiveness of the sector. In addition to focusing on short term growth strategies, a strategic shift towards longer term capability building strategies such as research and development and human capital and skills development was required.

Extensive literature also support that companies that continuously invest in R&D to build technological capabilities attain stable growth and strong financial performance. The most innovative organizations globally leverage on fostering a culture of research and development, innovation and continuous learning. In addition, the most innovative companies were leveraging on strong intellectual property, organizational growth and strategic flexibility. These levers were clearly lacking in most companies surveyed, and as such a clear strategic intent towards acquiring these attributes was required in the medium to long term.

The local companies were clearly failing to utilize the knowledge based in public research and development institutions. Publicly funded academic and R&D institutions are rich sources of innovative ideas. Such institutions act as cheap conduits for the requisite knowledge imperative for building the technological and innovation activities in the iron and steel sector. Instead of relying on private consultants and commercial R&D laboratories, the companies could tap into the local knowledge base available in most state universities, vocational colleges and public research institutions. With the right collaboration, such research institutions could be instrumental in spearheading participation in futuristic scientific research and development of new products and processes, especially in the new and emerging iron and steel making processes that are being developed globally to leverage on low cost of manufacturing and raw material flexibility.

With the cyclical swings currently being experienced in intermediate iron and steel products, the full value chain participation towards stainless steel refining is imperative. Through private-public partnerships, priority towards establishing a stainless steel refining plant is in the best interest on the country. Again, the role of public, private and social partnerships play a critical role in the success of such capital intensive business ventures.

Zimbabwe lacks a holistic beneficiation framework to incentivize the value addition and beneficiation of iron and steel products. There is need for a clear policy framework to integrate the iron and steel sector into the broader value addition and beneficiation framework as espoused in the ZIM ASSET economic blueprint. In addition, it is important to take into account the technological and innovation gaps through a through cost-benefit and SWOT analyses of the iron and steel sector before crafting such a beneficiation framework. Since the iron and steel industry does not operate in isolation, Zimbabwe also needs an economy-wide national innovation surveys to determine the gaps and linkages from broader economy perspectives.

Conclusions

This paper evaluated the technical, managerial and organizational capabilities of the companies in the iron and steel industry in Zimbabwe. The results show identified technological and innovation capability gaps that were affecting the competitiveness of the companies in the sector. Finally, the paper made recommendations to integrate the sector specific technological and innovation best practices into the broader national beneficiation framework.

References

1. Adler, P. a. C. K., 1991. Behind the Learning Curve: ASketch of the Learning Process. *Management Science,* Volume 37, pp. 267-281.

2. Ahuja, G. a. L. C., 2001. Entreprenuership in the Large Corporation: A Longitudinal Study of How Established Firms Create Breakthrough Inventions. *Strategic Management Journal,* 22(6-7), pp. 521-543.

3. AMG Vanadium, 2009. *Ferroalloys and Alloying Additives Online Handbook.* [Online] Available at: http://www.metallurgvanadium.com/chromiumpage.html [Accessed 22 April 2014].

4. Andrew, J. H. K. M. D. S. H. T. A., 2009. *Boston Consulting Group.* [Online] Available at: www.bcg.com [Accessed 22 February 2014].

5. Aranda, D. a. M.-F. L., 2002. Determinants of Innovation through a Knowledge-based Theory Lens. *Industrial Management and Data Systems,* 102(5), pp. 289-296.

6. Australian Government Initiative, undated. *Innovation.* [Online] Available at: http://www.business.gov.au/BusinessTopics/Innovation/Pages/Whatisinnovation.aspx [Accessed 19 February 2014].

7. Australian Institute for Commercialization, 2009. *Why Innovate?.* [Online] Available at: http://www.innovationtoolbox.com.au/why-innovate/innovation-can-be-incremental-or-radical [Accessed 12 April 2014].

8. Barney, J., 2001. Is the Resource Based View a Useful Perspective for Strategic Management Research? Yes. *Academy of Management Review,* 26(1), pp. 41-56.

9. Baxter, R., 2005. *Beneficiation: Mining Industry ViewPoint.* Johannesburg, LBMA, pp. 25-28.

10. Bell, A., 2013. *Zimbabwe Steel Industry Facing Collapse Amid Worsening Investment Climate.* [Online] Available at: http://www.thezimbabwean.co/news/zimbabwe/65371/zim-steel-industry-facing-collapse.html [Accessed 2014 April 2014].

11. Bessant, J. a. T. J., 2007. *Innovation and Entreprenuership.* s.l.:John Wiley and Sons.

12. Bessant, J. a. T. J. P. K., 2013. *Managing Innovation: Integrating Technological Market and Organizational Change.* 5th ed. s.l.:Wiley.

13. Bessant, J. a. T. J. P. K., 2013. *Managing Innovation: Integrating Technological Market and Organizational Change, 5th Ed.* s.l.:Wiley.

14. Booz and Company, 2005. *Relationship between R&D Spending and Sales Growth Earnings or Shareholder Returns,* NY: Booz and Comoany.

15. Branscomb, L. a. A. P., 2002. *Between Invention and Innovation: An Analysis of Funding for Early-Stage Technology Development.* [Online] Available at: http://belfercenter.hks.harvard.edu/publication/2067/between_invention_and_innovation.html [Accessed 27 January 2014].

16. Burns, N. a. G. S., 2003. *Understanding Nursing Research.* 3rd ed. Philadelphia, PA: W. B. Saunders.

17. Business and IP Centre, undated. *Qualitative and Quantitative Research.* [Online] Available at: http://www.bl.uk/bipc/resmark/qualquantresearch/qualquantresearch.html [Accessed 3 May 2014].

18. Cassel, C. a. S. G., 1994. Qualitative Research in Work Contexts. In: *Qualitative Methods in organizational Research: A Practical Guide.* London: Sage, pp. 1-13.

19. Cavaye, A., 1996. Case Study Research: A Multi-facted Research Approach for IS. *Information Systems Journal,* Volume 6, pp. 227-242.

20. Center for Accelerating Innovation, 2006. *Innovation Metrics: Measurement to Insight,* Washington: George Washington University.

21. Chesbrough, H., 2003. The Era of Open Innovation. *Sloan Management Review,* 44(3), pp. 35-41.

22. Chiesa, V. a. M. C., 1996. Searching for an Effective Measure of R&D performance. *Management Decision,* 34(7), pp. 49-57.

23. Confederation of Zimbabwean Industries, 2013. *Manufacturing Survey, 2013,* Harare: Confederation of Zimbabwean Industry.

24. Cooper, D. a. S. P., 2008. *Business Research Methods.* 10th ed. Boston: McGraw Hill.

25. Cooper, R., 2007. Managing Technology Development Projects. *IEEE Engineering Management Review,* p. 67.

26. Crawford, M. a. B. D., 2002. *New product Development.* 7th ed. Boston: McGraw Hill.

27. CZI, 2013. *Manufacturing Survey, 2013,* Harare: Confederation of Zimbabwean Industry.

28. Deloitte, 2011. *Positioning for Mineral Beneficiation: Opportunity Knocks,* Johannesburg: Deloitte.

29. Department of Mineral Resources, 2011. *Minerals Resources, Republic of South Africa.* [Online] Available at: http://www.dmr.gov.za/beneficiation-economics.html [Accessed 18 February 2014].

30. Department of Science and Technology, 2002. *South Africa's National Research and Development Strategy,* Johannesburg: Republic of South Africa.

31. Detecon Consulting, 2013. *Innovation Performance Measurement- Assessing and Driving the Innovation performance of Comapnies.* [Online] Available at: http://innovationcenter.deteconusa.com/articles/innovation-performance-measurement-assessing-and-driving-the-innovation-performance-of-companies/ [Accessed 19 April 2014].

32. Elliot, M. A. A. A. C. S. B. M. P. a. B. A., 2013. *Global Steel 2013: A New World, A New Strategy,* New York: Ernst and Young.

33. Ernst and Young, 2013. *Africa 2013: Getting Down to Business,* Johannesburg: Ernst and Young.

34. Ernst and Young, 2013. *Cost Control and Margin protection in the South African Mining and Metals Industry,* Johannesburg: Ernst and Young.

35. European Commission, 2006. *Handbook of Recommended Practices for Questionnaire Development and Testing in the European Statistical System,* s.l.: European Commission.

36. EuroStat, 2013. *Community Innovation Survey.* [Online] Available at: http://epp.eurostat.ec.europa.eu/portal/page/portal/microdata/cis [Accessed 25 April 2014].

37. France, C. M. C. a. W. D. a., undated. *The Innovation Imperatives: How Leaders can Build an Innovation Engine.* [Online] Available at: https://www.mmc.com/knowledgecenter/viewpoint/The_Innovation_Imperative.php [Accessed 19 April 2014].

38. Gabriel, D., 2013. *Inductive and Deductive Approaches to Research.* [Online Available at: http://deborahgabriel.com/2013/03/17/inductive-and-deductive-approaches-to-research/ [Accessed 30 April 2014].

39. General Management Review, 2005. *The Key to Adapting and Succeeding in any Environment.* [Online] Available at: http://www.etgmr.com/jan_mar05/Innovation.html [Accessed 30 March 2014].

40. Gerybadze, A. 2010. R&D, Innovation and Growth: Performance of the World's Leading Corporations. In: *Innovation and International Corporate Growth.* Berlin: Springer-Verlag Berlin Heidelberg, pp. 11-29.

41. Government of Zimbabwe, 2014. *ZIM ASSET Policy Document,* Harare: Government of Zimbabwe.

42. Government of Zimbabwe, 2014. *ZIMASSET Policy Document,* Harare: Government of Zimbabwe.

43. Hansen, M. a. B. J., 2007. *The Innovation Value Chain,* Harvard: Havard Business Review.

44. Harada, T. a. T. H., 2011. Future Steel Making Model by Direct Reduction. *Iron and Steel Institute of Japan,* p. 1301.

45. Harada, T. a. T. H., 2011. Future Steel Making Model by Direct Reduction. *Iron and Steel Institute of Japan,* p. 1301.

46. Harder, H., 2010. Explanatory Case Study. In: *Encyclopedia of Case Study Research.* Thousand Oaks, CA: Sage, pp. 371-372.

47. Henderson, B., undated. *The Product Portfolio (Reprint).* [Online] Available at: https://www.bcgperspectives.com/content/Classics/strategy_the_product_portfolio/ [Accessed 24 April 2014].

48. Hosen, Z. N. M. a. N. F., 2011. Efficiency Measurement of Capacity Utilization in Pharmaceutical Industry. *Cost and Management,* Volume January-February, pp. 25-28.

49. Huh, K., 2010. Steel Consumption and Economic Growth in Korea: Long Term and Short-tem Evidence. *Resource Policy,* 36(2), pp. 107-113.

50. Human Resources Research Council, 2008. *South African Innovation Survey Main Results 2008,* Johannesburg: Department of Science and Technology (ZA).

51. International Stainless Steel Forum, 2013. *ISSF Annual Review 2013,* s.l.: http://www.worldstainless.org/Files/ISSF/non-image-files/PDF/ISSF_Annual_Review_2013.pdf.

52. Ireland, R. K. D. a. C. J., 2003. Antecedents, Elements and Concequences of Corporate Entrepreneurship Strategy. *Academy of Management Proceedings,* pp. L1-6.

53. Jaruzelski, B. M. C. N. R. R. a. S. H., 2012. Role of Coherent Linkgaes in Fostering Innovation Based Economies in The Gulf Cooperation Council Counries. In: *The Global Innovation 2012.* s.l.:World Intellectual Property Organization, pp. 109-120.

54. Jick, T. D., 1979. Mixing Qualitative and Quantitative Methods: Traiangulation in Action. *Adminstrative Science Quarterly,* Volume 24, pp. 602-611.

55. Johnson, R. and. O. A., 2004. Mixed Methods Research: A Research Paradigm Whose Time Has Come. *Educational Researcher,* 33(7), pp. 14-26.

56. Jourdan, P. C. G. K. I. a. C. E., 2012. *Mining Sector Study.* [Online] Available at: http://www.zeparu.co.zw [Accessed 26 April 2014].

57. Kaplan R.S., N. D. a. R. B., 2010. Managing Alliances with the Balanced Scorecard. *Harvard Business Review,* 88(1), pp. 114-120.

58. Kaplan, R. a. N. D., 1996. The Balanced Scorecard: Translating Strategy into Action. *Harvard Business School Press.*

59. Kaplan, R. a. N. D., 2004. Strategy Maps: Converting Intangible Assets into Tangible Outcomes. *Harvard Business School Press.*

60. Kaplan, S., undated. *Measuring Innovation to Drive Business Growth: The Complete Guide to Innovation Metrics.* [Online] Available at: http://www.innovation-point.com/innovationmetrics.htm [Accessed 14 April 2014].

61. Khomba, J. V. F. a. G. D., 2011. Redesigning an Innovation Section of the Balanced Scorecard Model: An African Perspective. *Southern African Business Review,* 15(3).

62. Kiron, D. K. N. H. K. R. M. a. G. E., 2013. *The Inovation Bottom Line.* [Online Available at: https://www.bcgperspectives.com/content/articles/sustainability_energy_environment_innovation_bottom_line/ [Accessed 18 February 2014].

63. Kohlbacher, F., 2005. The Use of Qualitative Content Analysis in Case Stusy Research. *Qualitative Social Research,* 7(1).

64. Kotter, J., 2012. *Key to Changing Organizational Culture.* [Online] Available at: http://www.forbes.com/sites/johnkotter/2012/09/27/the-key-to-changing-organizational-culture/ [Accessed 30 April 2014].

65. KPMG, 2012. *Global Manufacturing Outlook: Fostering Growth through Innovation,* New York: KPMG Global.

66. Kumba Iron Ore, 2011. *The South African Iron and Steel Value Chain,* Johannesburg: AngloAmerican.

67. Lall, S., 1992. Technological Capabilities and Industrialization. *Wordl Development,* 20(2), pp. 165-186.

68. Lall, S., 1992. Technological Capabilities and Industrialization. *Wordl Development,* 20(2), pp. 165-186.

69. Lall, S. a. p. C., 2004. Failing to Compete: Technology Development and Technology Ssystems in Africa. *Journal of African Economics,* Volume 13, pp. 195-198.

70. Lawson, B. a. S. D., 2001. Developing Innovation Capabilities in Organizations: A Dynamic Capabilities Approach. *International Journal of Innovation Management,* 5(3), pp. 377-400.

71. Lawson, B. a. S. D., 2001. Developing Innovation Capability in Organizations: A Dynamic Capabilities Approach. *International Journal of Innovation Management,* 5(3), pp. 377-400.

72. Löfsten, H., 2014. Product Innovation Processes and Trade-off Between product Innovation Performance and Business Performance. *European Journal of Innovation Management,* 17(1), pp. 61-84.

73. Lynn, D., 1991. *The Application of Case Study Evaluations.* [Online] Available at: http://PAREonline.net/getvn.asp?v=2 and n=9 [Accessed 4 May 2014].

74. Malerba, S., 1992. Learning by Firsm and Incremental Technical Change. *The Economic Journal,* Volume 102, pp. 845-859.

75. Malinoski, M. a. P. G., 2011. *How Do I Measure Innovation?*, North Carolina: The Balanced Scorecard Institute.

76. Mallampally, P. and. S. K., 1999. *Foreign Direct Investment in Developing Countries*. [Online] Available at: http://www.imf.org/external/pubs/ft/fandd/1999/03/mallampa.htm [Accessed 22 April 2014].

77. Manning, C. a. F. R. J., 2001. Emerging Technologies for Iron and Steel. *Journal of Materials*, 53(10), pp. 20-23.

78. Martowidjojo, A. a. A. F., 2011. *The Role of Organizational Learning on Innovation Value Chain*. s.l., Business Innovation and Technology Management, pp. 113-116.

79. Matinde, E. H. J. M. T. M. L. a. K. I., 2014. *Beneficiation Conference 2014*. [Online] Available at: http://www.chamberofminesofzimbabwe.com/publications/category/19-benefaction-conference-2014.html [Accessed 22 April 2014].

80. Matinde, E. H. J. M. T. M. L. a. K. I., 2014. *Beneficiation Conference 2014*. [Online] Available at: http://www.chamberofminesofzimbabwe.com/publications/category/19-benefaction-conference-2014.html [Accessed 22 April 2014].

81. Matinde, E. N. L. C. E. a. C. G., in press. *Zimbabwe's Engineering and Metals Sector Value Chain Analysis*, Harare: SIRDC-ZEPARU-USAID Collaborative Study.

82. Mayring, P., 2001. Combination and Integration of Quantitative and Qualitative Analysis. *Forum: Qualitative Social Research Sozialforschung*, 2(1).

83. McAdam, R. R. R. a. S. M., 2014. Determinants for Innovation Implementation at SME and Inter SME levels within Peripheral Regions. *International Journal of Entreprnuerial Behavior and Research*, 20(1), pp. 66-90.

84. McGregor, J., 2004. *Are patents a Measure of Innovation*. [Online] Available at: http://www.businessweek.com/stories/2007-05-04/are-patents-the-measure-of-innovation-businessweek-business-news-stock-market-and-financial-advice [Accessed 25 April 2014].

85. Mikkola, J., 2001. Portfolio Management of R&D Projects: Implications for Innovation Management. *Technovation*, Volume 21, pp. 423-435.

86. Ministry of Industry and Commerce, 2011. *Zimbabwe Industrial Development Policy 2011-2015*, Harare: Government of Zimbabwe.

87. Ministry of Science and Technology Development, 2012. *Zimbabwe Nanotechnology Statement*, Harare: Government of Zimbabwe.

88. Ministry of Science and Technology, 2012. *Second Science, Technology and Innovation Policy*, Harare: Government of Zimbabwe.

89. MMCZ, undated. *Chrome*. [Online] Available at: http://www.mmcz.co.zw/index.php?option=com_content and view=article and id=63:metals-products and catid=42:metals [Accessed 22 April 2014].

90. Muller, A. V. L. a. M. P., 2005. Metrics for Innovation: Guidelines for Developing a Customized Suite of Innovation Metrics. *Strategy per cent leadership*, 33(1), pp. 37-45.

91. Neely, A. F. R. F. C. V. A. a. H. J., 2001. Framework for Analyzing Business performance, Firm Innovation and Related Contextual Factors: Perceptions of Managers and Policy Makers in two European Regins. *Integrated Manufacturing Systems*, 12(2), pp. 114-124.

92. Neely, A. G. M. a. P. K., 2005. Performonace Measurement System Design: A Literature Review and Research Agenda. *Internationa Journal of Operations and Production Management*, 25(12), pp. 1228-1263.

93. Neely, A. M. J. P. K. H. G. M. B. M. a. K. M., 2000. Peformonace Measurement System Design: Developing and Testing a Proces-based Approach. *International Journal of Operations and Production Management*, 20(10), pp. 1119-1145.

94. Nicolaides, A., 2014. Research and Innovation-The Drivers of Economic Development. *African Journal of Hospitality, Tourism and Leisure*, 3(2), pp. 1-16.

95. O'Connor, C., 2010. Product Innovation and Management. In: *Wiley International Encyclopedia of Marketing*. NY: John Wiley and Sons.

96. OECD, 2002. *Frascatti manual Measurement of Scientific and Technological Activities*, Paris: OECD.

97. OECD, 2005. *Oslo Manual: Guidelines for Cllecting and Interpretting Innovation Data*, Oslo: OECD.

98. OECD, 2005. *Oslo Manual: Guidelines for Collecting and Interpretting Innovation Data*, Oslo: OECD.

99. OECD, 2011. *Glossary of Statistical Terms*. [Online] Available at: https://stats. oecd.org/glossary/detail.asp?ID=1322 [Accessed 12 April 2014].

100. Outokumpu, 2012. *Stainless Steel Market Review*. [Online] Available at: http:// ar2011.outokumpu.com/business/stainless-steel-market-review [Accessed 22 April 2014].

101. Outokumpu, 2012. *Stainless Steel Review*. [Online] Available at: http://reports. outokumpu.com/en/2012/business/our-operating-environment/market-review/ [Accessed 22 April 2014].

102. Parker, J., 2014. *The Potential for Stainless Steel in Zimbabwe*. Victoria Falls, Chamber of Mines Zimbabwe.

103. Parliament of Zimbabwe, 2013. *Chrome Mining Sector in Zimbabwe*, Harare: Government of Zimbabwe.

104. Pietrobelli, C., 2006. *Fostering Technological Capabilities in Sub-Saharan Africa*. [Online] Available at: http://www.scidev.net/global/policy-brief/fostering-technological-capabilities-in-sub-sahara.html [Accessed 24 April 2014].

105. Porter, M., 1985. *Competitive Strategy*. New York: Free Press/Mcmillan.

106. Price Waterhouse Coopers, 2012. *How to Drive Innovation and Business Growth: Leveraging Emerging Technology for Sustainable Growth*, New York: PriceWaterhouseCoopers.

107. Price Waterhouse Coopers, 2013. *Breakthrough Innovation and Growth.* [Online] Available at: http://www.pwc.com/gx/en/innovationsurvey/index.jhtml [Accessed 12 4 2014].

108. PwC, 2013. *Innovation, growth and Business Strategy,* London: Price Waterhouse Coopers.

109. Robson, C., 2002. *Real World Research.* Oxford: Blackwell.

110. Ro, S., 2013. *Business Insider.* [Online] Available at: http://www.businessinsider.com/rd-correlated-to-stock-price-returns-2013-8 [Accessed 14 February 2014].

111. sagepub.com, undated. *Chapter 2: Research Philosophy and Qualitative Interviews.* [Online] Available at: http://www.sagepub.com/upm-data/43179_2.pdf [Accessed 29 April 2014].

112. Sanchez, J. a. P. H., 2012. Innovation: Case Study Among Wood, Enenergy and Medical Firms. *Business Process Management Journal,* 18(6), pp. 898-918.

113. Saunders, M. L. P. T. A., 2009. *Research Methods for Business Students.* 5 ed. Essex: Pearson Education.

114. Saunila, M. a. U. J., 2013. Facilitating Innovation Capability through Performance Measurement: A Study of Finnish SMEs. *Management Research Review,* 36(10), pp. 991-1010.

115. Schumpeter, J., 1912. *The Theory of Economic Development: An Inquiry into Profits, Capital, Credit, Interest, and the Business Cycle.* Cambridge: Harvard University Press.

116. Schumpeter, J., 1934. *The Theory of Economic Development: An Inquiry into Profits, Capital, Credit, Interest, and the Business Cycle.* Cambridge: Harvard University Press.

117. Schumpeter, J., 1939. *Business Cycles: A Theoretical, Historical, and Statistical Analysis of the Capitalist Process.* New York: McGraw-Hill.

118. Sheilds, L. a. T. A., 2003. The Difference Bewteen Quantitative and Qualitative Research: Research Update. *Paediatric Nursing,* 15(9), p. 24.

119. Slocum, A. a. R. E., 2008. *Understanding Radical Technology Innovation and its Applicaion to CO2 Capture,* Pittsburg: Carnegie Institute of Technology.

120. Tashakkori, A. and. T. C., 1998. *Mixed Methodlogy: Combining Qualitative and Quantitative Approaches.* Thousand Oaks: Sage.

121. Taticchi, P. B. K. a. T. F., 2012. Performance Measurement and Management Systems: State of the Art, Guidelines for Design and Challenges. *Measuring Business Execellence,* 16(2), pp. 41-54.

122. Technologies, U. N. U. I. f. N., 2004. *Designing a Policy-Relevant Innovation Survey for NEPAD,* Maastricht: UNU-INTECH.

123. Thurmond, V., 2001. The Point of Traingulation. *Journal of Nursing Scholarship,* pp. 254-256.

124. Total Materia, 2008. *Production of Stainless Steel.* [Online] Available at: http://www.keytometals.com/page.aspx?ID=CheckArticle and site=kts and NM=220 [Accessed 23 March 2014].

125. Trainor, K. K. M. a. A. R., 2013. Effects of Relational Proclivity and Marketing Intelligence on New Product Development. *Marketing Intelligence and Planning,* 31(7), pp. 788-806.

126. Tseng, M. L. S. a. V. T., 2012. Mediate effect of Technology Innovation Capabilities Investment Capabilities and Firm Performance. *Proceedia-Social and Behavioral Sciences,* Volume 40, pp. 817-829.

127. Tushman, M. a. O. C., 1996. Abidextrous Organizations: Managing Evolutionary and Revolutionary Change. *California Management Review,* 38(4), pp. 8-29.

128. UNCTAD, 2007. *The Least Developed Countried Report: Knowledge, Technological learning and Innovation for Development,* geneva/New York: United Nations Conference on Trade and Development.

129. UNECA, 2013. *Outcome Statement on the Ad Hoc Experts Group Meeting on Industrization for Economic Transformation and Sustainable Development,* Harare: United Nations Economic Commission for Africa Southern Africa.

130. United Nations Economic and Social Council, undated. *Science, Technology and Innovation.* [Online] Available at: http://www.un.org/en/ecosoc/about/science.shtml [Accessed 14 February 2014].

131. USGS, 2012. *Nickel,* NY: United States Geological Survey.

132. Voelpel, S. L. M. E. R. D. T., 2005. *The Tyranny of the Balanced Scorecard in the Innovation Economy.* Cambridge, Cambridge University.

133. Wagner, K. T. A. Z. H. a. F. E., 2013. *The Most Innovative Companies in 2013: Lessons From Leaders,* Boston: Boston Consulting Group.

134. World Bank, 2013. *Tanzania Enterprise Survey.* [Online] Available at: http://microdata.worldbank.org/index.php/catalog/2023/study-description [Accessed 26 2014 July].

135. World Intellectual Property Organization, undated. *Patents.* [Online] Available at: http://www.wipo.int/patents/en/ [Accessed 14 April 2014].

136. World Steel Association, 2014. *Crude Steel Statistics Data 2013.* [Online] Available at: http://www.worldsteel.org/media-centre/press-releases/2014/World-crude-steel-output-increases-by-3-5—in-2013.html [Accessed 2014 April 2014].

137. Wortler, M. S. F. H. R. P. H. a. V. N., 2010. *Flexibility and Innovation: Today's Imperatives for Steel,* Boston: Boston Consulting Group.

138. Yam, R. G. J. P. K. a. T. E., 2004. An Audit of Technological Innovation Capabiities in Chinse Firms: Some Emprirical Findings in Beijing, China. *Research Policy,* 33(8), pp. 1123-1140.

139. Yin, R., 2003. *Case Study Research: Design and Methods.* Carlifornia: Sage Publications.

140. Zhu, Y., 2008. *Sectoral Study on the Iron and Steel Industry: Interdependence on Energy and Climate Security,* London: Chatham House.

141. Zimbabwe Statistics Office, 2012. Harare: Government of Zimbabwe.

142. Zoz, K. a. L. G., 1997. *Patents and R&D: An Econometric Investigation using Applications for German, European and US patents by German Comanies,* GmBH: Center for European Economic Research.

Chapter 3

The Role of Zimbabwe Republic Police Minerals and Border Control Unit (MBCU) in the Mining Sector

Rabson Mpofu

Senior Assistant Commissioner,
Officer Commanding Minerals and Border Control Unit
E-mail: rkampofu@gmail.com

ABSTRACT

The paper presents the role of the Zimbabwe Republic Police [ZRP], particularly the Minerals and Border Control Unit in ensuring enduring and purposeful safeguarding of mineral resources in Zimbabwe. The role of the ZRP transcends mere enforcement of the law to ensure a conducive business environment where players in the mining, agricultural, tourism and manufacturing sectors, among others, religiously hold fast to ethics governing their conduct.

Introduction

It is a pleasure for me to make a presentation at this important occasion, of Non-Aligned Movement Science and Technology Conference on minerals processing and Mineral Beneficiation. Allow me, first and foremost, to welcome all our esteemed NAM dignitaries to this very beautiful and peaceful nation of Zimbabwe. I am convinced that all the delegates present here have so far enjoyed their stay in this country and have thus witnessed firsthand, the peace and tranquility prevailing in the country.

Zimbabwe is a peace loving nation, quite hospitable, and as the law enforcement agents in the country, we would like to assure you that your safety and security during your stay is indeed guaranteed. As the Zimbabwe Republic Police we always ensure that peace, which is without doubt a vital ingredient of successful and sustainable investment, prevails all the times. Hopefully, after this conference, some of you shall have the opportunity to visit places of interest and have the practical feel of the entrenched peace and hospitality of the people of Zimbabwe.

Scope of the Presentation

As part of our input to the Non–Aligned Movement Science and Technology Conference on minerals processing and beneficiation, this presentation serves to expose the role of the Zimbabwe Republic Police [ZRP], particularly the Minerals and Border Control Unit in ensuring enduring and purposeful safeguarding of mineral resources in Zimbabwe. The role of the ZRP transcends mere enforcement of the law to ensure a conducive business environment where players in the mining, agricultural, tourism and manufacturing sectors, among others, religiously hold fast to ethics governing their conduct. Accordingly, before we even dwell into mineral beneficiation, the basic and underlying requisite is to put in place sound mechanisms to safeguard these minerals.

This conference therefore gives the Zimbabwe Republic Police the platform to elaborate the role the police force plays in the mining sector through its arm, the Minerals and Border Control Unit. To this end, the scope of my presentation shall focus on the following pertinent issues, among others:

☆ The role of the ZRP Minerals and Border Control Unit,

☆ Statutes empowering the police in its operational endeavors in the mining sector,

☆ The prevailing rapport between the police and mining entities and other stakeholders,

☆ Milestones by the Minerals Unit's policing endeavors of the mining sector,

☆ The Conference Theme from the ZRP perspective,

☆ The way forward for NAM member countries from a National Security paradigm.

The Role of the ZRP MBCU

☆ Zimbabwe as a country is blessed with numerous minerals which range from gold, diamonds, platinum, emeralds chrome among others.

☆ As expected, this has attracted unscrupulous syndicates, who in cohort with the locals, are siphoning or smuggling the country's minerals beyond our borders.

It is against this backdrop that in its wisdom, the Government of Zimbabwe formulated strategies designed to regulate both the exploitation as well as the disposal of these minerals.

There are countries in this world that are known to have rich mineral deposits, but these resources have become a bane more than a blessing as the vultures of this world busied themselves looting at the expense of the citizens.

This then culminated in the establishment of ZRP, MBCU in the year 2006. Originally the ZRP had the Gold Squad Unit which concentrated on safeguarding smuggling of gold; hence MBCU in essence is the gold squad which was transformed into a more expanded entity with some broadened responsibilities.

Thus, the MBCU's mandate includes the following, among others;

1. Curbing illegal exploitation, dealings and smuggling of high value minerals [gold and diamonds],
2. Arresting and causing the prosecution of all those found on the wrong side of the law with regard to the mining and trading of the said minerals,
3. Enforcement of legislation defining and regulating mining activities in Zimbabwe,
4. Manning all entry and exit ports of the country with the view to detecting potential smuggling of precious minerals,
5. Guaranteeing the security of mining concerns that deal in precious minerals by deploying overtly and covertly at gold and diamond mining enterprises,
6. Securing yet to be exploited concerns rich in precious mineral deposits, and
7. Providing motorized and armed escorts during the transportation of minerals from source to storage warehouses.

The Enabling Legislation

The Zimbabwe Republic Police's involvement in mining activities in the country is governed by several enabling pieces of legislation. In essence, our being in the sector is above board. Some of the statutes include:

1. **The Gold Trade Act, Chapter 21:03:** This piece of legislation prescribes as well as proscribes certain activities when dealing in and possession of gold. The said ACT also enunciates the issuance of licenses and permits to those interested in mining the precious metal. It is the prerogative of MBCU to ensure that the dictates of the Gold Trade Act are adhered to. NB [Gold, just like water, is ubiquitous in Zimbabwe]
2. **Precious Stones Trade Act, Chapter 21:06:** Ladies and gentlemen, the availability of precious stones in Zimbabwe prompted the legislature to come up with the Precious Stones Act. Its purpose is to address the following:
   ☆ Unlawful dealing in or possession of precious stones
   ☆ Conditions on which licensed dealers or permit holders may deal in or posses precious stones

☆ Conditions on which miners may deal in or posses precious stones

☆ Issuance and or cancellation of licenses/permits

☆ Procedure on confiscation of precious stones

☆ Offences and penalties thereof.

It is thus the prerogative of the law enforcement agents, in this case, the MBCU to ensure that the ACT is religiously adhered to without any equivocation.

3. **Mines and Minerals [Custom Milling Regulations] Chapter 21:05:** The regulations are meant to govern operations of Custom Milling Plant operators as they process ores on behalf of miners. To plug any possible leakages the Custom Milling Plant operators are permitted to buy all gold realised from their mills and subsequently sell to Fidelity Printers and Refiners.

4. **Explosives Act 10.08:** The mining industry in Zimbabwe extensively utilizes explosives when blasting rock ores. It is against this background that the police monitor the use of explosives in the mining sector with a view of curbing the misuse of explosives by syndicates engaging in illegal mining and other nefarious activities. Whenever the abuse of explosives is detected, the MBCU will deal with such malcontents.

5. **Environmental Management Act, Chapter 20.27:** Mineral exploitation is heavily associated with environmental degradation. Often, both legal and illegal miners give little regard to the environment as they go about their mining businesses. The Environmental Management Agency [EMA], a body mandated with the protection of the environment normally ramps in the police in dealing with miners who disregard the provisions of the law. The apprehension and bringing to court of those found wanting and those that are illegally mining and dealing with precious minerals and causing land degradation becomes the duty of the police. In this regard the MBCU is working with the Environmental Management Authority striving to bring those elements bent on defiling the environment to court.

6. **Minerals Marketing Corporation of Zimbabwe, Act Chapter 21.04:** Part IV of the said Act prescribes and proscribes the sale and production of minerals among other issues. The MBCU is thus mandated to keep an eye on the sale of precious minerals in particular with the view to curbing all such transactions that are not done within the precincts of the law.

7. **Customs and Excise Act: Chapter 23.02:** Some illegal fortune seekers from within and outside Zimbabwe's borders have been a menace since the discovery of alluvial diamonds at Chiadzwa Diamond Field in Manicaland Province. Syndicates bent on siphoning the precious stones to some far away destinations have been on the prowl and trying to illegally export their loot. In this regard, the MBCU in conjunction with other stakeholders have been on high alert over the years on a mission to burst such unscrupulous entities. Within the country's ports of entry and exit, searches of vehicles and luggage on transit have led to the seizure of

precious minerals being smuggled out of Zimbabwe. The Customs and Excise Act has always been handy in this regard.

Multi-Sectoral Approach

☆ The Zimbabwe Republic Police enjoys cordial relations with the Ministry of Mines and Mining Development, the mining sector and other players.

☆ As police, we cannot effectively enforce mining legislation without constant collaboration with all players in the mining sector.

☆ Apart from constantly liaising with MBCU on operational issues, the Ministry of Mines and Mining Development also assists police in numerous ways.

☆ The Zimbabwe Republic Police emphasizes the multi-sectoral approach in policing and safeguarding of minerals.

Milestones Made So Far

☆ All official exit and entry points across the country's borders are now manned by members of MBCU to curb smuggling of precious minerals.

☆ There has been substantial improvement in gold being delivered to Fidelity of Printers and Refiners.

☆ Police have managed to attain normalcy in the diamond industry by ensuring orderliness in Chiadzwa diamond mining fields after the initial rush following the discovery of diamond deposits which threatened to destroy the environment and cause illegal trade in the precious stones.

☆ ZRP has managed to forge sound partnerships with stakeholders in the mining sector including members of the public in its concerted efforts to curb smuggling of minerals.

☆ Police managed to establish MBCU structures across the country.

☆ ZRP conducts constant training of the members of MBCU in conjunction with experts in the mining sector.

Policing and Beneficiation

☆ Zimbabwe Republic Police like any other player in Zimbabwe stands to benefit from mineral processing and beneficiation.

☆ Mineral beneficiation creates employment and this goes a long way in providing a means of earning a living to people who would have otherwise turned to criminality as a source of earning a living.

☆ Mineral beneficiation can eradicate poverty and reduce civil unrests caused by hunger.

☆ With enhanced economic growth, the government can increase its fiscal support to law enforcement agencies thereby buttressing their effectiveness.

☆ Improved economy and remuneration of members can also curb corruption.

☆ In short, mineral beneficiation enhances national security.

Way Forwards

☆ Given the benefits of mineral processing and beneficiation to the economy and national security, NAM member states and other developing countries should assist each other in coming up with robust measures to enhance beneficiation.

☆ Platforms of this nature and magnitude should be conducted regularly since they provide for the sharing of important ideas.

Conclusions

I would like to thank you for affording me this platform to discuss with you the role being played by the Zimbabwe Republic Police in safeguarding mineral resources in the country. We remain hopeful that the ideas you have shared will help our nation to strengthen our economy and in turn bolster our operations as a police organisation.

Chapter 4

Minerals and Mineralogy in Mauritius

R. Goodary

Head of Geotechnical Laboratory and Dean,
Faculty of Sustainable Development and Engineering,
Université des Mascareignes, Mauritius
E-mail: rgoodary@udm.ac.mu

ABSTRACT

Mauritius, an island located in the Indian Ocean with 1.2 million population has almost negligible mineral resource as such, but however its vision and know how make the country become visible among countries engaged in minerals processing and beneficiation. This paper aims at highlighting the country's activities and its future endeavours in the field of mineralogy, the act that governs the extraction and exploitation of minerals and finally the constitution of the soils and other formations in the island. The main mineral activity is associated with diamond cutting and polishing industry for the purpose of exportation and this brings a total revenue 2.6 billion rupees which is commendable for a country without a least grain of diamond. X ray diffractograms revealing the minerals present in local soils have been analysed and the presence of two main components, namely: Kaolinite and Montmorillonite are evidenced.

Keywords: Minerals and mineralogy, Ocean economy, Soil weathering, Kaolinite, Montmorillonite.

Introduction

Mauritius has no mineral resources as such. However, the mining and quarrying activity, which in the context of Mauritius covers salt production; and stone and sand quarries are reflected in the annual statistical reports of Statistics Mauritius. In 2013, 24 establishments were involved in this sector with total employment of 1,020. Gross output stood at Rs 2,930m. The share of this industry in the Gross Domestic product is negligible.

Above all these, basalt stone is the main exploited mineral in the country, being used for various purposes ranging from crafts souvenirs to building blocks.

The use of minerals in Mauritius is also associated with the diamond cutting and polishing industry. In line with the export oriented strategy, the first diamond cutting and polishing factory was set up in the 1970's. Over the years, the sector has witnessed considerable development to the extent that Mauritius has become a reliable sub-contracting centre for this highly closed shop business. At present, 4 enterprises mostly foreign-owned are involved in cutting and polishing of diamonds in Mauritius. The biggest of these enterprises is a subsidiary of Tiffany, the world renowned US multinational jeweler. The finished diamonds and precious stones are exported mainly to the USA, Belgium, Vietnam, Switzerland, Hong Kong and France. They employ a total workforce of around 900 persons and realized exports of about Rs 2.6 billion in 2013 contributing to around 5 per cent of total manufacturing exports. These companies import rough diamonds from countries like Canada, Australia, Russia, Tanzania, Belgium and Sierra Leone. All imports of rough diamonds are covered by the Kimberley Certificate of origin certifying that they are from legal sources.

Capacity Building

On the job-training at factories has helped significantly to develop a pool of highly skilled and experienced diamond cutters and polishers. In addition, the School of Jewellery operated by Mauritius Institute for Training and Development (MITD) has played a crucial role to provide basic training to workers who were more apt to join the industry. Government has also set up a Gemology Laboratory at the Assay Office for testing of precious and semi-precious stones. Four officers have been trained in gemology in India to effectively run the lab.

Vision of Mauritius for an Ocean Economy

In the context of the vision of Mauritius to promote an ocean economy, consideration is being given to the extraction of minerals from the sea. One of the first priorities will be to develop a Petroleum and Minerals Exploration and Extraction Framework Agreement (Budget Speech 2014). This part is being largely developed below (information gathered from the website http://www.oceaneconomy.mu).

Mauritius manages a maritime zone of 2.3 million km^2 with a large geographic territory and the competencies, technologies and systems to manage its territory. The potential for economic advancement and prosperity that this resource can generate, if developed in a sustainable way, could take Mauritius to a high-income country as inspired by the Prime Minister in his recent speech.

As on date the ocean territory of the island contributes significantly to the wealth of Mauritius. The GDP share was estimated at 10.8 per cent in 2012, with an added value of 32.5 billion Mauritian Rupees, of which over 90 per cent came from three established sectors – coastal tourism and marine leisure, seaport-related and seafood-related activities where significant scope for future growth exists. Mauritius now endeavours to exploit the ocean and expected resources from new clusters. Among the economic activities identified as possible priority areas are the following: the utilization of pure, nutrient-rich and cold deep sea water to develop Deep Ocean Water Application (DOWA) projects which will provide sea-water air conditioning to industrial and commercial users, reducing the dependence of Mauritius on fossil fuels, as well as create a new business for aquaculture and related activities like seaweed and algal culture, cosmetics and pharmaceuticals, agrochemicals, water bottling and thalassotherapy, among others. As such, two firms who have shown interest for extracting this water are already being considered, one in the city and a second in the south of the island. In the same light, marine renewable energies can in the long term considerably reduce our dependency on fossil fuels.

The mapping and stock-taking of the seabed for both living organisms and potential hydrocarbon and mineral resources in our waters is another promising area. The discovery of hydrocarbons in our EEZ would be a potential game changer for the Mauritian economy. The granitic nature of the Seychelles Islands and the discovery of a thick sedimentary sequence in the Seychelles plateau have attracted oil companies to prospect in the region. Recent geophysical surveys in the region of the Mascarene Plateau revealed that the continental crust along the Mascarene plateau extend further southward to the Banks. Already, the discovery in 2009 of inactive hydrothermal fields by a joint Mauritian and Japanese expedition within our EEZ indicates the likelihood of mineral deposits. In fact, prospecting nations have recently requested and been allocated deep sea mining blocks by the International Seabed Authority in areas contiguous to the Mauritian EEZ.

The Minerals Act of Mauritius (1966)

The Minerals Act of the country regulates all mineralogical and related aspects linked with identification, extraction and exploitation of minerals (The constitution of Mauritius).

The act minerals include:

- ☆ Firstly metaliferous minerals containing aluminium, antimony, arsenic, barium, bismuth, cadmium, cerium, chromium, cobalt, colombium, copper, iron, lead, lithium, magnesium, manganese, mercury, molybdenum, nickel, potassium, sodium, tantalum, tin, titanium, tungsten, uranium, vanadium, zinc, zironium, and all other substances of a similar nature to any of them, and all ores containing them and combinations of any of them with each other or with any other substance, other than those occurring in the form of precious minerals;

- ☆ Secondly combustible carbonaceous minerals including coal; lignite, which includes brown coal and any coal which the President may prescribe to be lignite;

- ☆ Thirdly are the other minerals, including those used for their abrasive or refractory qualities and asbestos, barytes, bauxite, china clay, crystals, fuller's earth, graphite, laterite, marble, mica, nitrates, pipeclay, potash, pumice, quartz, slate, soda, sulphur, talc, and all other substances of a similar nature to any of them; and

☆ Fourthly are the precious minerals, including -

    (a) precious stones and semi-precious stones including amber, amethyst, beryl, cat's eye, chrysolite, garnet, and all other semi-precious stones, whether of the same kind as those enumerated or not;

    (b) precious metals;

But does not include:

    (i) pottery, clay or rock salt;

    (ii) any materials, such as clay, sand, limestone, sandstone, or other stones, commonly used for the purpose of road making, building or for the manufacture of any article used in the construction of buildings where such material does not contain any valuable metal or precious stone;

    (iii) petroleum and associated substances as defined in the Petroleum Act; "Minister" means the Minister to whom responsibility for the subject of commerce and industry is assigned; "precious stones" means diamonds, emeralds, opals, rubies, sapphires, turquoises, and such other stones as may be prescribed to be precious stones for the purposes of this Act; "prospect" means search for minerals and includes such working as is reasonably necessary to enable the prospector to test the mineral-bearing qualities of the land.

Right to Prospect for Minerals

Subject to the Act, and until such time as the President may by regulations prescribe, no person shall prospect for, mine or work minerals in or under any land in Mauritius whether he is the owner of the land or not.

Exclusive Right of Government

The Government shall have the exclusive right to prospect for minerals in or under any land. The Minister may authorize in writing any person to carry on prospecting operations in or under any land on behalf of the Government.

Notification of Intention to Prospect

☆ The Minister shall by notice published in the Gazette and in 3 daily newspapers give notice of the intention of Government to prospect for minerals in or under any land.

☆ A notice under first subsection shall give a brief description of the land.

Powers of Authorised Prospector

On the issue of a notice under section 6, the authorised prospector may -

    (a) Enter upon the land in respect of which the notice has been given with his workmen;

    (b) Employ in prospecting on the land any number of persons;

    (c) Do all other acts necessary to prospect for minerals in the land.

Compensation

The Government shall pay compensation to the owner or occupier of the land in or under which prospecting operations are carried out for any:

(a) Disturbance of the rights of the owner or occupier;

(b) Damage done to the surface of the land; or

(c) Damage caused to any crops, trees, buildings or works on the land.

The President may make such regulations as he thinks fit for the purposes of this Act.

The Geology of Mauritius

The volcanic islands of Mauritius are located about 800 km east of Madagascar in the Indian Ocean. Mauritius consists of the main island of Mauritius, the much smaller island of Rodrigues and two smaller groups of islands and reefs to the north and northeast of the main island. All islands are of volcanic origin and are surrounded by coral reefs.

The mineral resource of Mauritius is negligible. The main minerals being quarried are basalts for construction purposes with smaller amounts of lime being produced from local coral limestone and coral sand. Potentially important are the polymetallic nodules that occur on the ocean floor at about 4,000 m depth around Mauritius. They contain more than 15 per cent of both iron and manganese and more than 0.35 per cent cobalt. Other potential 'agro minerals' available in this island country are crushed basaltic rocks, calcareous coral sands, and raised coral reef deposits.

The forests of Mauritius are small in area but perform vital functions, the most important of them being soil and water conservation. Mauritius has no known oil, natural gas or coal reserves, and therefore depends on imported petroleum products to meet most of its energy requirements.

It consists of undulating central uplands (rising to a maximum elevation of about 600min the south and with a mean elevation of the order of 400m) surrounded by mountain ranges and plains, forming a bowl with chipped rims, filled with layers of young formations excess of which have eroded away. Outside are the plains, which were deposited as lava flows. These flows are the products of small volcanoes situated on the wide low median ridge running across the island from southwest to north east. The much eroded relics of the rim protrude above these younger volcanic regions as discontinuous ring mountain ranges with rugged peaks. The mountain ranges surrounding the central plateau have an asymmetrical profile.

The formation of Mauritius Island consists of four volcanic series (Figure 4.1) as follows (Proag, 1995 and Giorgi *et al.*, 1999):

☆ Emergence

☆ Older volcanic series Early lavas

☆ Intermediate or Early volcanic series

☆ Younger volcanic series early lavas or Late lavas.

(E.S.W. SIMPSON 1951)

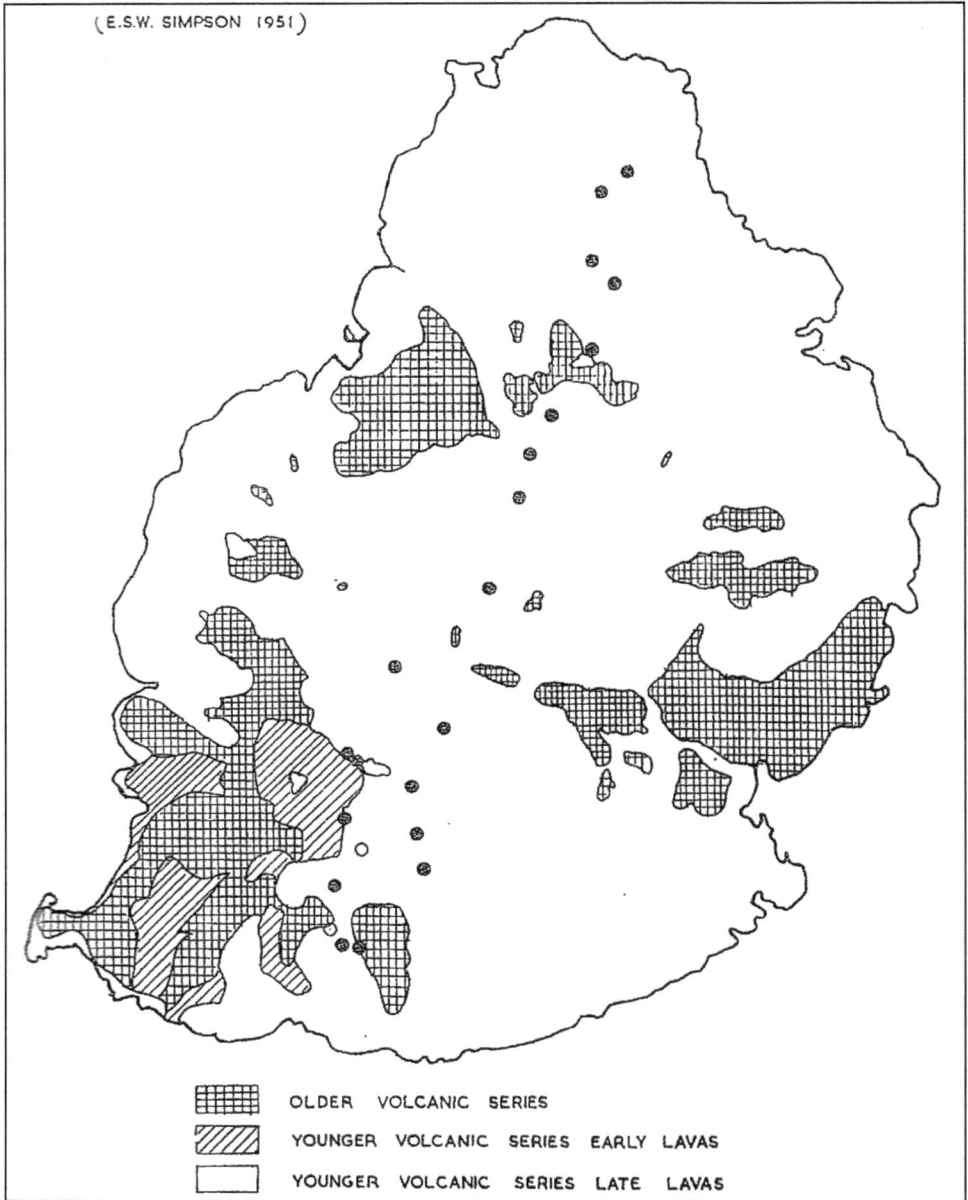

OLDER   VOLCANIC   SERIES

YOUNGER   VOLCANIC   SERIES   EARLY   LAVAS

YOUNGER   VOLCANIC   SERIES   LATE   LAVAS

Figure 4.1: Geological Map of Mauritius

The lava flows consist of sequence of massive basalt strata and volcanic breccias, the residual soils in Mauritius are the end product of severe weathering and leaching of basalt, belonging to the family of laterite and lateric soils covering much of the tropics, subtropics and warm temperate regions of the earth.

Minerals Classification

The recent lava flows have given rise to rocky soils in Mauritius, developing throughout the full range climatic conditions. As the rocky soils are developing on easily weatherable, highly basaltic lavas, they present many of the profile characteristics of their neighbouring soils derived from older lava flows and yet they are generally shallow, rocky and non-eroded.The weathering of these formations have given rise to various classes of soils with typical mineralogies.

The Low Humic Latosols

The low humic latosols have developed on the intermediate lavas which occur in the sub-humid lower rainfall humid zones with a distinct dry season. The clay fraction is composed of kaolin cemented with oxides, they have, in field, the texture of silty clays or silty clay loams, even though mechanical analysis gives up to 80 per cent of clay content. The physical properties of the low humic latosols are on the whole good, although in the drier areas they tend to be compact and somewhat sticky when wet.

The low humic latosols have been divided into four families which shows differences resulting from variation in rainfall and age of the parent material

Humic Latosols

The humic latosols have developed on the intermediate lavas in the humid to super-humid areas. The soils are mainly clays, but behave like silty clays, and they do not show any textural differentiation of horizons.

The humic latosol group has been divided into two families:

☆ Rosalie –transitional to the low Humic Latosols

☆ Riche Bois – transitional to the Humic Ferruginous Latosol

Humic Ferruginous Latosol

It is a strongly weathered soils of the high rainfall region of the island. These soils occur on the Early and Intermediate Lavas of the Younger Volcanic series and on the lavas of the older volcanic series, and therefore range from mature soils with only a few iron oxide concretions to almost senile soils with large amounts of iron oxide and bauxitic concretions.

The property of these soils is, naturally, the amount of iron oxides and alumina as concretions in the top part of the profile. The amount of concretions present indicates the state of weathering, in other words they represent an integration of the intensity of weathering (Shootenko *et al.*, 1988 and 1989).

It have been divided into four families:

☆ Chamarel – typical to the early lavas and ash deposit from younger volcanic series

☆ Belle Rive

☆ Sans Souci

☆ Midlands

Latosolic Soils

The latosolic soils are intrazonal soils developing on the late lavas, which on the surface at least are being highly vesiculated and fractured. The soils developing on the range from neutral soils with high basicity to strongly acid soils with only a small content of exchangeable bases.

All the soils have varying quantities of boulders and gravels of fresh or weathered basalt in their profiles and they all have a high content of organic matter. Rock outcrops are frequent in many areas, their extent giving rise to phases of rockiness. The rockiness have been based on the amount of bed-rock exposure, but when bulldozed this reflects the amounts of rocks.

Latosolic Reddish Prairie Soils

The whole sample contains varying amounts of cobbles and gravels of unweathered basalt and bed-rock exposures are common. The soils are slightly acid to neutral. The soils vary in texture from clay loams to silty clays.

It has been divided into three groups:

☆ Medine (weathered moderately, mostly overlying older soils)

☆ Labourdonnais(weathered moderately, shallow)

☆ Mont choisy (slightly weathered, shallow)

Latosolic Brown Forest Soils

They have been formed on the late lavas in the super-humid area. They will develop with age into humic ferruginous latosols and in fact in some areas so much iron oxide deposition has occurred that they approach very closely to being zonal soils. They contain varying amounts of weathered gravels and cobbles within the sample, due to the high rainfall they have been strongly leached of bases.

These minerals are unique in Mauritius which have a good impact of liming to cane growth.it is comparatively high in exchangeable manganese and in organic matter. The soils vary in texture from clay loams to silty clays or clay.

Many small areas form part of this group of mineral. It has been divided into two group according to the weathering intensity:

☆ Rose belle (rainfall 2000mm-4000mm/yr)

☆ Bois chéri (rainfall >3000mm/yr)

Dark Magnesium Clay

A residual basalt of volcanic origin belonging to the lateritc soil group and coal fly ash are both considered to be unsuitable for construction purposes. It consists of minerals with physical properties showing the dominant influence with a high amount of saturation by magnesium. The main spot of these soils has developed under low rainfall influence. They are self-mulching black or dark grey clays, the lower levels of which show accumulations of calcium carbonate, with or without gypsum crystals below the carbonate level.

The dark magnesium clays present all the agricultural problems of an intractable clay, they are plastic when wet and very hard when dry, and they need drainage in the wet season and irrigation during the dry season; in addition the high quantity of magnesium.

It has been divided into two groups:

☆ Plaine lauzun-(presence of gypsum crystal)

☆ Magenta

Hydromorphic Soils

This type of mineral has been formed in the presence of excess of water in the sample during part of the year. They occur in all the rainfall areas and at all point of view on flat land or in depressions where drainage is sub-optimal.

This type of soil has been divide into two minerals:

Grey Gydromorphic Soils

These minerals are plastic clays produced as a result of periodical water logging. It has been divided into two groups:

☆ Balaclava (strongly hydromorphic)

☆ St. Andre (moderately hydromorphic)

The chemical composition of the two groups are similar, the are neutral, have cation exchange and base saturation is 45 per cent. The minerals are high in exchangeable magnesium.

An example of X-ray diffractograms of two typical soils, denoted LM1 and LM2, are shown below (Goodary *et al.*, 2012):

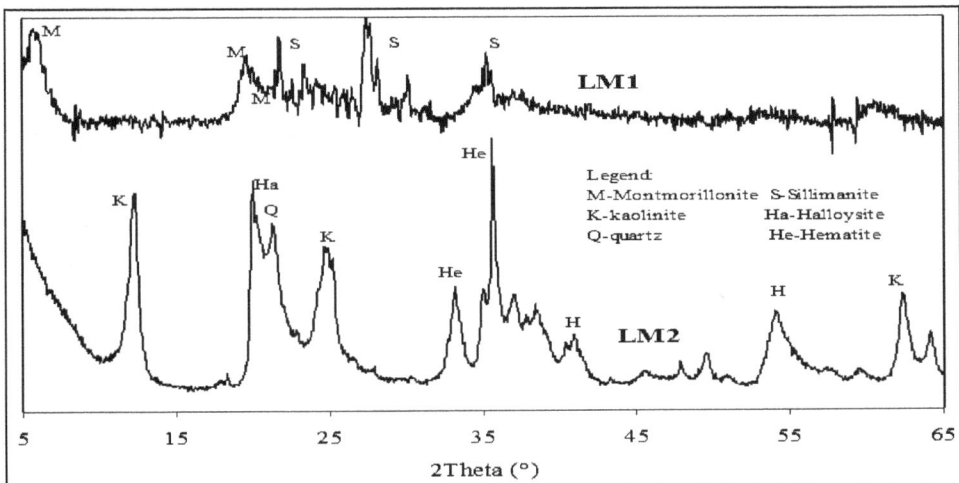

Figure 4.2: X-Ray Diffractograms of Two Typical Soils of Mauritius

Results of the above analysis (carried out at research laboratory based in Limoges, France), among others, give a clear indication of the mineralogy of existing subsoil which can be of great interest to specialists of the region.

Conclusions

1. The use of minerals in Mauritius is also associated with the diamond cutting and polishing industry, in line with the export oriented strategy.

2. The potential for economic advancement and prosperity that the Mauritian Ocean Economy strategy can generate, if developed in a sustainable way can take Mauritius to a high-income country.

3. The minerals act, treating all aspects of minerals namely, their extraction, exploitation and beneficiation has been elaborated.

4. A detailed geological history of Mauritius has been developed.

5. X ray diffractograms revealing the minerals present in local soils have been analysed and the presence to two main components namely Kaolinite and Montmorillonite are evidenced.

References

1. Proag V. The geology and water resources of Mauritius. Mahatma Gandhi Institute; Mauritius; 1995.

2. Carte Geologique et Hydrogeologique, Ile Maurice. Republique de Maurice; 1996.

3. The Constitution of Mauritius, The attorney General's office, Mauritius.

4. Giorgi L. and Borchiellini S. Les aquifères de l'Ile Maurice (Map); 1999.

5. British Standards Institution BS1377-2. Methods of tests for soils for Civil Engineering purposes. Part 2; Classification tests; 1990.

6. Goodary, R, Lecomte-Nana, G.L, Petit, C. and Smith, D. 2012. Investigation of the strength development in cement - stabilised soils of volcanic origin. Construction and Building Materials, 28: 592-598.

7. Shootenko, L.N. and Goodary, R. 1989. Effect of Degree of Weathering of Lateritic Soils on their Geotechnical Properties. Rehabilitation of Buildings, Kharkov, 47- 49.

8. Shootenko, L.N. and Goodary, R. 1988. A Statistical Analysis of Physical and Mechanical Properties of Lateritic Soils. Quality Amelioration in Urban Construction, Kharkov, 45- 52.

Chapter 5

Utilization of
Mineral Resources in Myanmar

Khin Maung Htwe

Associate Professor,
Department of Mining Engineering,
Technological University (Taunggyi), Myanmar
E-mail: khinmaunghtwe.mtu@gmail.com

ABSTRACT

Myanmar with a long and notable mining history in South-East Asia is a country rich in natural resources. It possesses extensive mineral resources and the most important among these are petroleum, natural gas, coal, tin-tungsten, lead, zinc, iron, copper, silver, gold, jade and gemstones. Union of Myanmar covers an area of about (676,580) sq. km within which there is some impressive mineralized areas and the mineral potential of the country is shown by widespread occurrences of mineral deposits. It was estimated earlier that there are about (583) sq.km shown tin-tungsten deposits, (50) sq.km of copper, (67) sq.km of lead-zinc, (45) sq.km of antimony and over (100) sq.km of industrial minerals including those of iron, nickel, chromium, manganese, coal, *etc.* In Myanmar, the current operation includes six coal mines (Kalewa, Namma, Maw Taung, Sanlaung, Lweje and Mya Ni). Myanmar is endowed with immense coal mine reserves which includes Kalewa coal mine -87 million tones, Namma coal mines 2 million tons, Tigyit coal mines-20 million tons, out of which 56,500 tons were exported in 1989.

The country has the highest quality rubies and jade in the world. Myanmar has world class deposits of lead-zinc-silver (Bawdwin mine) but is now decreasing the ore potential, tin-tungsten (Mawchi mine), fine rubies and sapphires (Mogok Stone Tract) and jade.

At present, the open-door economic policy of the government has attracted increased number of buyers from abroad. The Extractive Industries Transparency Initiative (EITI) leading authority was formed by a presidential order on December 14 in 2012. As a Consequence, Myanmar has applied for EITI membership and now been accepted as Candidate Country by International EITI Board.

Introduction

The union of Myanmar, covering an area of 676,580 square kilometers lies between about 10° N and 28° 30¢ N and 92° 30¢ E and 101° 30¢ E. The greatest north-south extent is about 2200 kilometer and the greatest east-west extent is about 950 kilometer. In the 800 kilometer long part of southern Myanmar, Tanintharyi area, the east-west extent is only 50 to 150 kilometer.

In the west, Myanmar is bordered by Bangladesh, in the northwest by India, in the north and northeast by China, in the east by the Loa People's Democratic Republic and in the southeast by Thailand. The total length of the country's borders is 4,000 km. The central and southern part of the country is bounded in the west by the 2,100 km long Rakhaine and Tanintharyi coastline of the Bay of Bengal and the Andaman Sea. The Union includes seven States and seven Regions.

Mineral Resource Base

Myanmar has a long history of mining of lead-zinc-copper, tin-tungsten, gold-silver, and gemstones. Myanmar has world class deposits such as the Bawdwin mine (the largest polymetallic base metal deposit in the world before the Second World War); but now decreasing of ore potential, the Mawchi mine (the largest tungsten-tin veins system before the Second World War) and the famous gemstone tract of the Mogok-Momeik area in northern Myanmar. The Government of Myanmar is now intensifying the exploration programme to expand the resource base of these deposits and country wide search for new deposits of these minerals. The total resource of the Monywa copper mine may exceed 2 billion tonnes of ore, giving an estimated mine life of more than 30 years.

Despite Myanmar's mineral wealth, the nation's mineral resources remain relatively under-exploited. While 86 per cent of the country has been geologically mapped, this mapping, and exploration, has not been particularly, undertaken in a major way because of poor equipment and a lack of secure access to parts of the country. The border regions, in particular, are not well explored. For these reasons, the exact magnitude of mineral reserves in Myanmar remains undetermined.

Utilization of Mineral Resources

Distribution and Reserves of Coal

The country is known to have numerous coal reserves, although many of them are of minor importance. According to the geological investigations completed up to now, there are 22 major coal deposits available throughout the country with estimated reserves of some 219 million tonnes. The distribution and reserves of coal, approximate analyses of coals in Myanmar are presented in Tables 5.1 and 5.2.

The major iron ore deposits in Myanmar are presented in Table 5.3.

Present Iron and Steel Production Activities

The No. 1 Iron and steel plant, the largest operational unit under the No.3 Mining Enterprise located at Anisakan, Pyin-Oo-Lwin Township, Mandalay Region.

The present iron and steel production activities, and metallic mineral deposits reserve in Myanmar are presented Tables 5.4 and 5.5.

Table 5.1: Distribution and Reserves of Coal

Sl.No.	Deposit	Locality	Reserve (Million Tons)	Category	Seam	Type of Coal
1	Kalewa	Kalewa Sagaing Div.	6.5	P-2	Upper	Sub-bituminous
			8	P-2	Lower	
			87	P-4		
2	Dar-Thwe-Kyauk	TamuSagaing Div.	33	P-3	–	Sub-bituminous
3	Pa-Lu-Zawa	KalewaSagaing	89	P-4	–	Sub-bituminous

Table 5.2: Approximate Analyses of Coals

Sl.No.	Description	Kalewa	Dar-thwe-kyauk	Pa-lu-zawa
1.	Ash	8.87 per cent	7.14 per cent	6.32 per cent
2.	Volatile Matter	38.67 per cent	40.36 per cent	42.00 per cent
3.	Fixed Carbon	52.50 per cent	47.54 per cent	46.00 per cent
4.	Sulphur	0.93 per cent	0.16 per cent	1.32 per cent
5.	Specific Gravity	1.35	1.33	1.30
6.	Calorific Value	11720 Btu/lb	12124 Btu/lb	11600 Btu/lb

Table 5.3: Major Iron Ore Deposits

Sl.No.	Deposit	Locality	Ore Reserve (Million tons)	Category	Fe Per cent	Type of Ore
1.	Kathaing Taung	Phakant Kachin State	223	P-2	50.54	Hematite 15% Magnetite 2% Goethite 75%
2.	La Maung	Phakant Kachin State	8.9	P-2	51.54	
3.	Peng Pet	Taungyi Shan State	10.7	P-2	56.5	Hematite
			60.0	P-2	42.6	Limonite
			10.0	P-2	43.2	Limonite
			29	P-2	–	Limonite
4.	Kyatwinye	Pyin-Oo-Lwin, Mandalay Region	3	P-2	54.0	Hematite 60% Magnetite 40%
5.	Kho-Kyun	Bokpyin, Taninthayi Region	3	P-2	56.0	
6.	Ma-Pu-Tae	Kawthaung, Taninthayi Region	1.3	P-2	54.0	
7.	Taungnyo Taung	Shwegu, Kachin State	18.9	P-4	40.67	Hematite Limonite

Table 5.4: Metallic Mineral Deposits Reserve

Sl.No.	Mineral Commodity	Occurrences	Deposits	Total	Million Tons				Remark
					P2	P3	P4	Total	
1.	Copper	88	27	115	1308.55	688.52	0.19	1997.3	
2.	Antimony	74	58	132	0.18	0.56	0.26	1.0	
3.	Chromite	23	20	43	0.02	0.04	0.02	0.08	
4.	Lead	163	128	291	13.28	12.67	18.14	44.09	
5.	Tin-Tungsten	125	358	483	0.36	38.74	0.29	39.9	Ore
					0.08	0.05	0.01	0.14	Conc.
6.	Nickel	4	10	14	70	80.66	12.2	162.86	
7.	Gold	246	95	341	15.29	47.66	3.16	66.11	Primary alluvial
					4.92	341.94	893.28	1240.2	(Cu-Yd)
8.	Iron	194	199	393	333	61.53	101.44	495.42	
9.	Zinc	18	11	29	5.21	14.41	0.33	19.95	

Table 5.5: Iron and Steel Industries in Myanmar

Sl.No.	Industry	Raw Material	Remark
1	No.1 Steel Industry, Aung Ian, Magwe Region	**Myingyan** and Import Billet	
2	No.2 Steel Industry Myaung Tagar, Yangon Region	Up To now Import	
3	No.3 Steel Industry Ywama, Yangon Region	Using scrap	
4	No.4 Steel Industry Myingyan, Mandalay Region	Using scrap and Lump Ore from Pinpet	DRP, Under construction
5	No.5 Steel Industry Pinpet, Southern Shan State	Pinpet, Southern Shan State Kathaing Taung, Kachine	

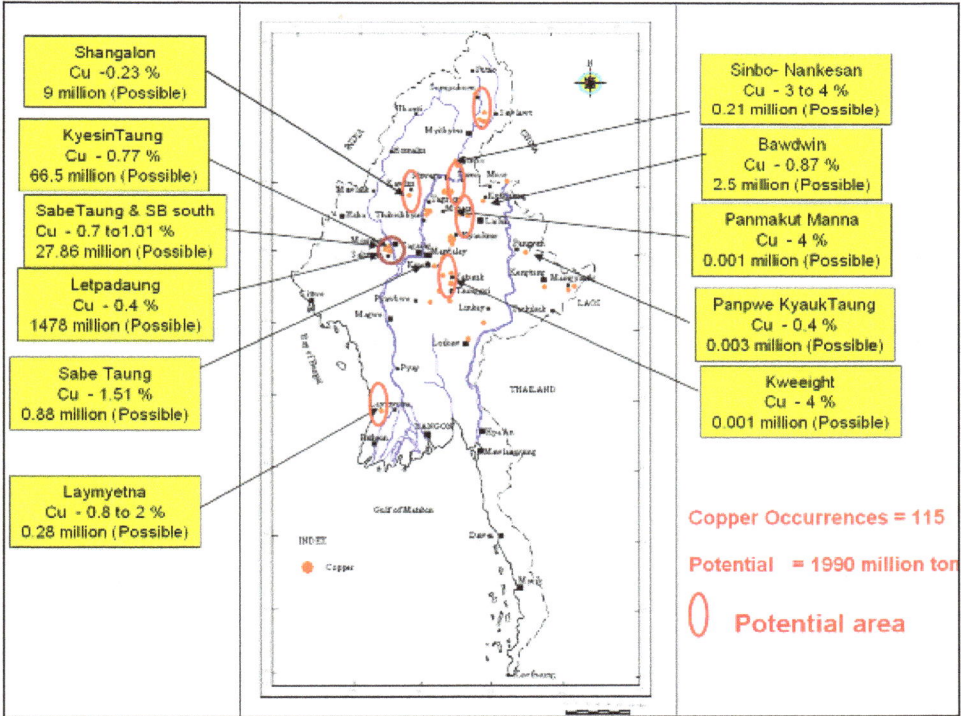

Figure 5.1: Distribution of Copper Deposits

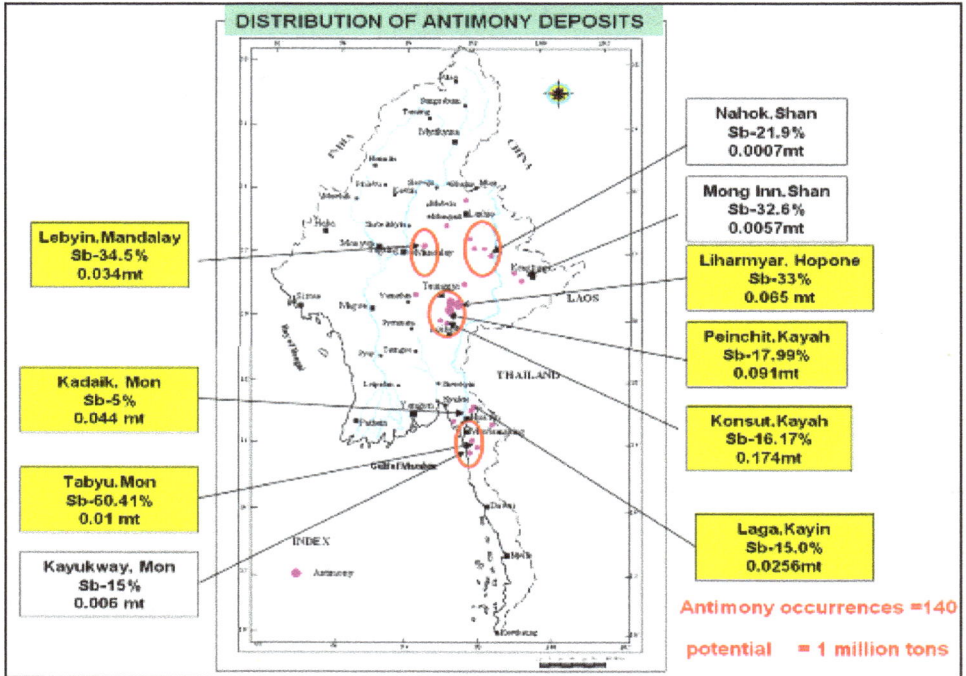

DISTRIBUTION OF ANTIMONY DEPOSITS

Nahok,Shan
Sb-21.9%
0.0007mt

Mong Inn,Shan
Sb-32.6%
0.0057mt

Liharmyar, Hopone
Sb-33%
0.065 mt

Peinchit,Kayah
Sb-17.99%
0.091mt

Konsut,Kayah
Sb-16.17%
0.174mt

Lebyin,Mandalay
Sb-34.5%
0.034mt

Kadaik, Mon
Sb-5%
0.044 mt

Tabyu,Mon
Sb-60.41%
0.01 mt

Kayukway, Mon
Sb-15%
0.006 mt

Laga,Kayin
Sb-15.0%
0.0256mt

Antimony occurrences =140
potential   = 1 million tons

Figure 5.2: Distribution of Antimony Deposits

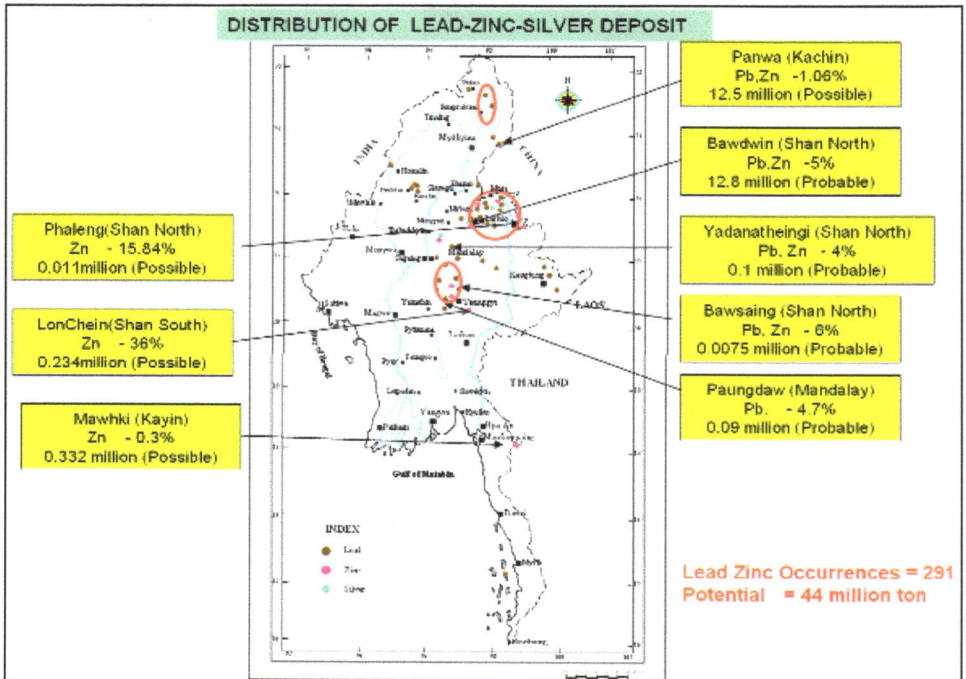

DISTRIBUTION OF LEAD-ZINC-SILVER DEPOSIT

Panwa (Kachin)
Pb,Zn -1.06%
12.5 million (Possible)

Bawdwin (Shan North)
Pb,Zn -5%
12.8 million (Probable)

Yadanatheingi (Shan North)
Pb. Zn - 4%
0.1 million (Probable)

Bawsaing (Shan North)
Pb. Zn - 6%
0.0075 million (Probable)

Paungdaw (Mandalay)
Pb. - 4.7%
0.09 million (Probable)

Phaleng(Shan North)
Zn   - 15.84%
0.011million (Possible)

LonChein(Shan South)
Zn   - 36%
0.234million (Possible)

Mawhki (Kayin)
Zn   - 0.3%
0.332 million (Possible)

Lead Zinc Occurrences = 291
Potential   = 44 million ton

Figure 5.3: Distribution of Lead-Zinc-Silver Deposits

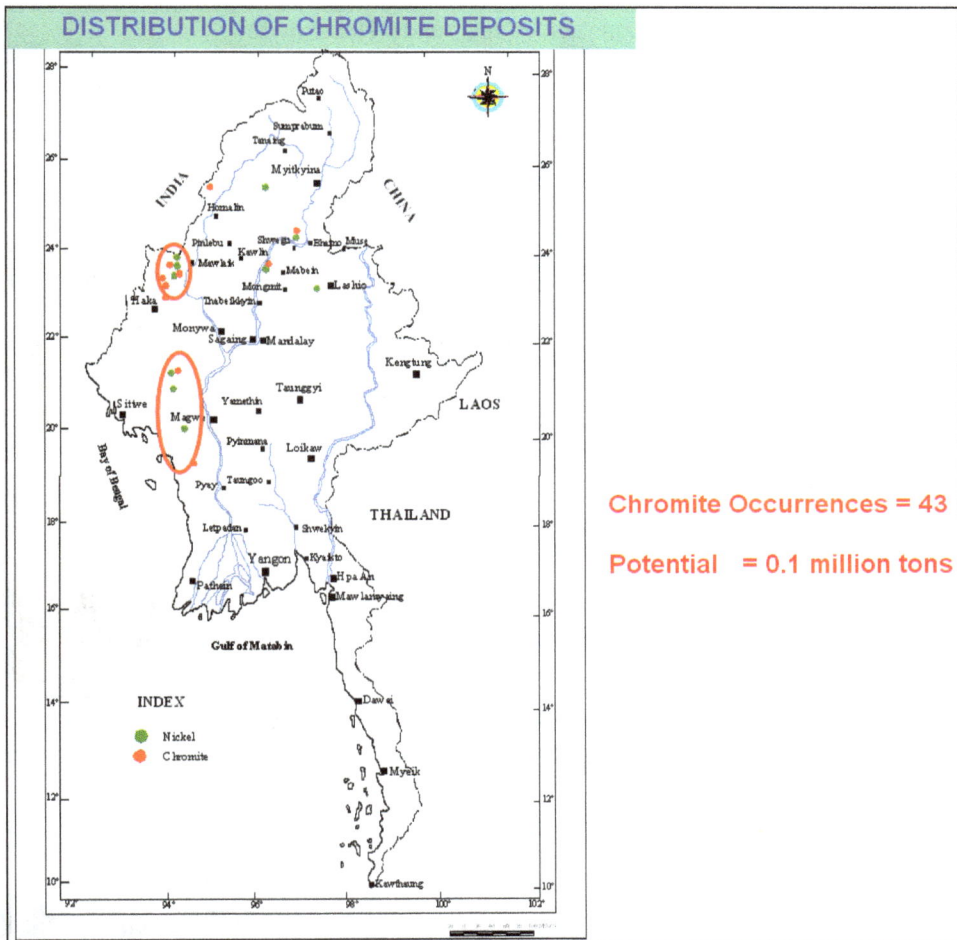

Figure 5.4: Distribution of Chromite Deposits

Mining, Benefication, Processing and Smelting in Myanmar

The mining, beneficiation, processing and smelting activities in Myanmar are shown in Table 5.6.

Table 5.6: Mining, Beneficiation, Processing and Smelting in Myanmar

Mining Methods	Beneficiation	Processing	Smelting
☆ Open pit Mining Method ☆ Underground Mining Method ☆ Placer Mining Method	☆ Crushing ☆ Grinding ☆ Gravity concentration flotation for producing concentrates	☆ Leaching and Electro-winning (Hydrometallurgy) SXEW, CIP and CIL	☆ Smelting (Pyrometallurgy) RKEF ☆ Combination of leaching and smelting

DISTRIBUTION OF TIN - TUNGSTEN DEPOSITS

Tin- Tungsten deposits= 480

Potential   = 40 million tons

Padatchaung (Primary)
Sn – 0.11%, WO₃ -0.81%
0.46 million (Probable)

Heinze (Placer)
Sn – 0.2- 0.3 lb/cu.yd.
0.012 million (Possible)

Kanbauk(Primary/ Placer)
Sn – 0.59%, 0.56 lb/cu.yd.
0.00865 million (Possible)

Atwin Bokpyin (Placer)
Sn – 0.56 lb/cu.yd.
0.0036 million (Probable)

Mawchi (Primary)
Sn – 0.32%
31 million (Probable)

Hermyingyi (Primary)
Sn – 0.37%
0.698 million (Probable)

Heinda (Placer)
Sn – 0.68 lb/cu.yd.
0.013 million (Probable)

KyaukmeTaung,
Pagaye(Placer)
Sn – 0.5 lb/cu.yd.
0.001 million (Probable)

Theindaw(Placer)
Sn – 0.36 lb/cu.yd.
0.0016 million (Probable)

Manawlon(Placer)
Sn – 0.6 lb/cu.yd.
0.0021 million (Probable)

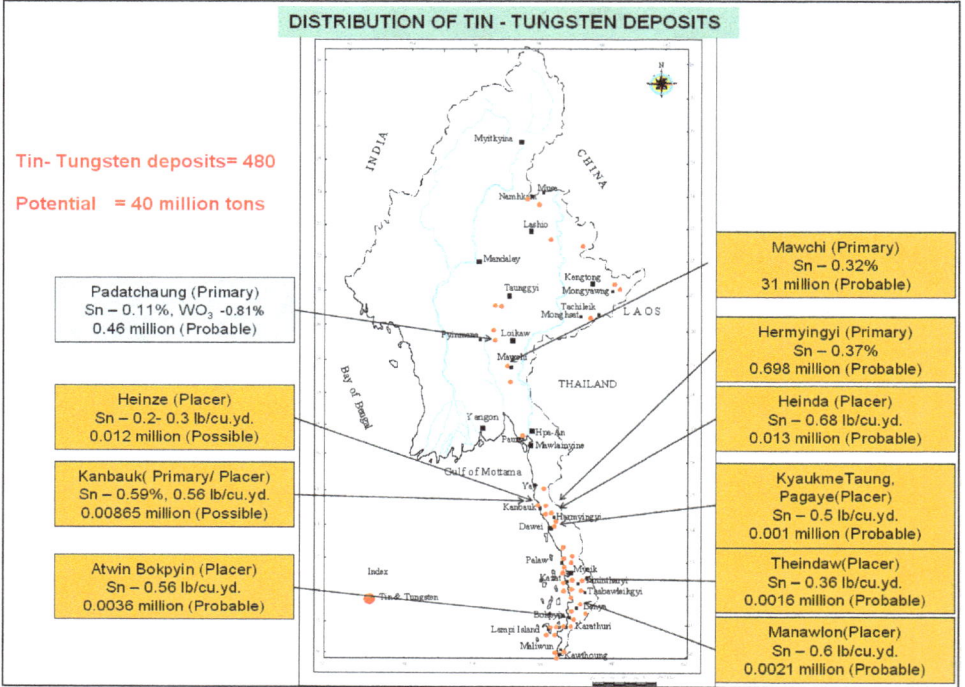

Figure 5.5: Distribution of Tin-Tungsten Deposits

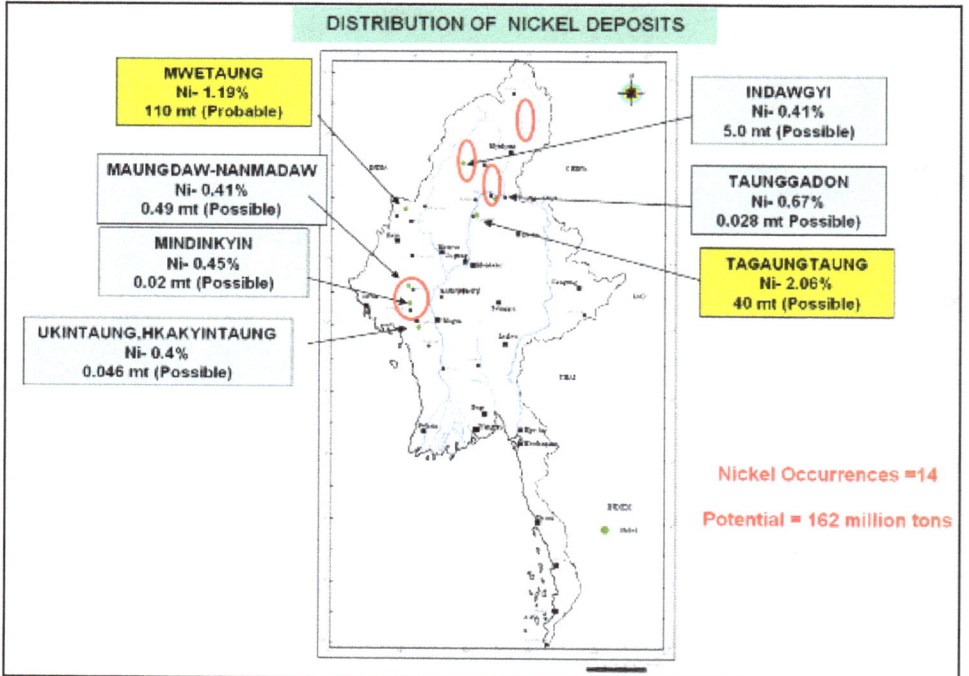

DISTRIBUTION OF NICKEL DEPOSITS

MWETAUNG
Ni- 1.19%
110 mt (Probable)

MAUNGDAW-NANMADAW
Ni- 0.41%
0.49 mt (Possible)

MINDINKYIN
Ni- 0.45%
0.02 mt (Possible)

UKINTAUNG,HKAKYINTAUNG
Ni- 0.4%
0.046 mt (Possible)

INDAWGYI
Ni- 0.41%
5.0 mt (Possible)

TAUNGGADON
Ni- 0.67%
0.028 mt Possible)

TAGAUNGTAUNG
Ni- 2.06%
40 mt (Possible)

Nickel Occurrences =14

Potential = 162 million tons

Figure 5.6: Distribution of Nickel Deposits

DISTRIBUTION OF GOLD- PLATINUM DEPOSITS

Shadusuik (Kachin)
Pt + Pd   - 0.01 gm/t
1 million (Possible)

Ngagyan (Kachin)
Pt + Pd   - 0.53%
21 million (Possible)

Shangalon (Sagaing)
Au   - 1.4-12 ppm
0.02 million (Possible)

Moehti Taung (Mandalay)
Au   - 15- 27 ppm
0.06 million (Probable)

Shwegyin (Bago)
Au   - 0.1-0.35 gm/t
1.2 million Cu.yd. (Probable)

Wakan- Tanaing (Kachin)
Au   - 0.04 gm/t
0.023 million Cu. Yd (Possible)

Namma- Kangon (Kachin)
Au - 0.13 gm/t
1.05 million Cu. Yd(Possible)

Banbwegon (Sagaing)
Au   - 3 ppm
6 million (Probable)

Kwinthonse (Mandalay)
Au   - 2-4 ppm
1.4 million (Probable)

Phayaungtaung (Mandalay)
Au   - 4 ppm
3.7 million (Probable)

Pyinmana (Mandalay)
Au   - 2 ppm
0.9 million (Possible)

Taunggu (Bago)
Au   - 0.2-0.5 gm/t
0.2 million (Possible)

INDEX

Gold (Primary)
Gold (Placer)
Platinum

Gold occurrences = 341

Potential   = 66 million tons

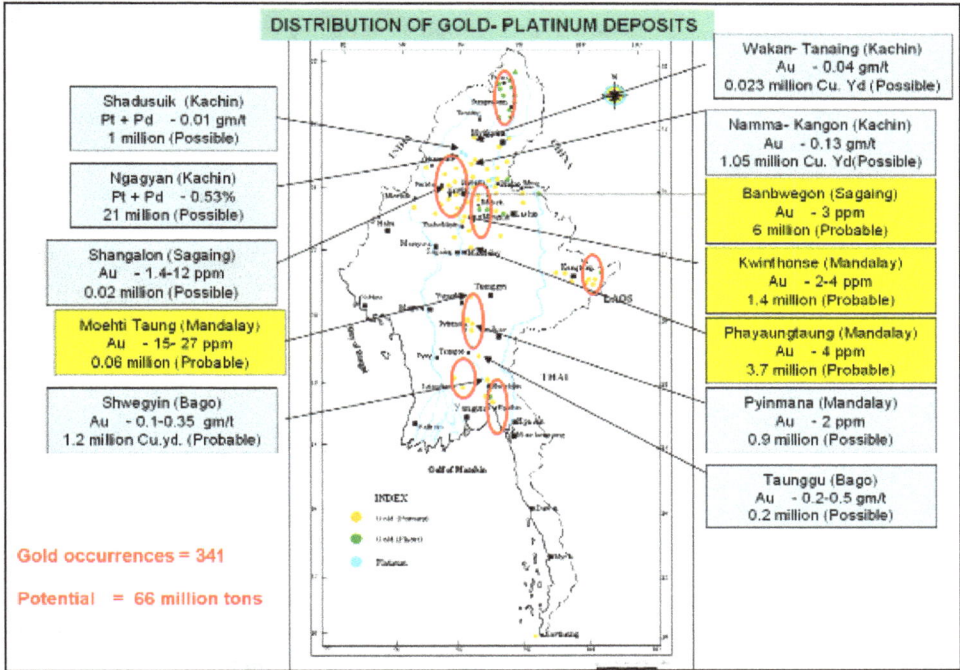

Figure 5.7: Distribution of Platinum Deposits

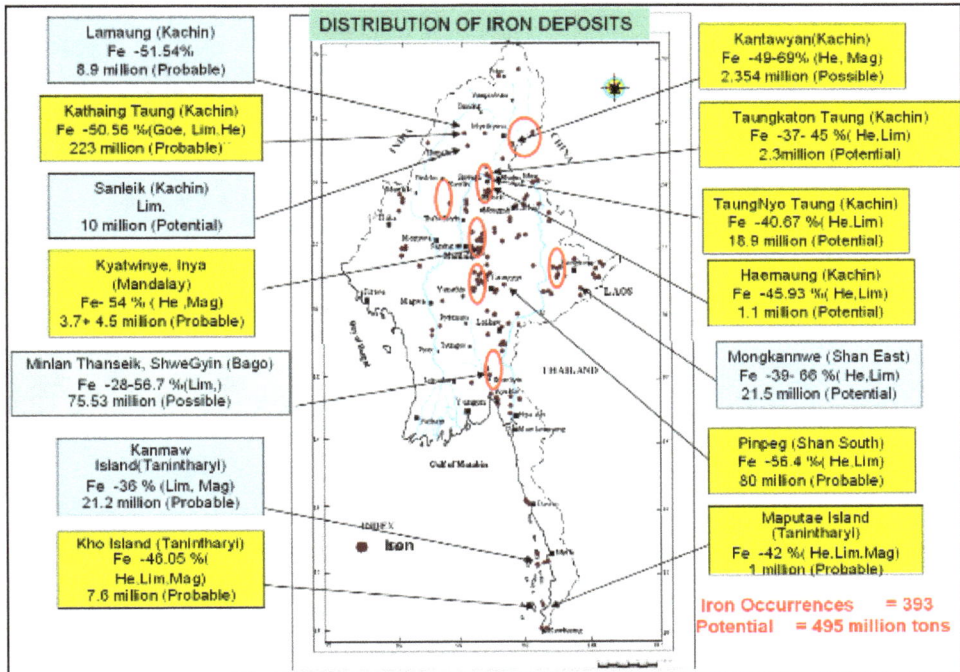

DISTRIBUTION OF IRON DEPOSITS

Lamaung (Kachin)
Fe -51.54%
8.9 million (Probable)

Kathaing Taung (Kachin)
Fe -50.56 %(Goe, Lim.He)
223 million (Probable)

Sanleik (Kachin)
Lim.
10 million (Potential)

Kyatwinye, Inya
(Mandalay)
Fe- 54 %(He ,Mag)
3.7+ 4.5 million (Probable)

Minlan Thanseik, ShweGyin (Bago)
Fe -28-56.7 %(Lim,)
75.53 million (Possible)

Kanmaw
Island(Tanintharyl)
Fe -36 % (Lim, Mag)
21.2 million (Probable)

Kho Island (Tanintharyl)
Fe -46.05 %(
He.Lim.Mag)
7.6 million (Probable)

Kantawyan(Kachin)
Fe -49-69%(He, Mag)
2,354 million (Possible)

Taungkaton Taung (Kachin)
Fe -37- 45 % (He.Lim)
2.3million (Potential)

TaungNyo Taung (Kachin)
Fe -40.67 %(He.Lim)
18.9 million (Potential)

Haemaung (Kachin)
Fe -45.93 %(He.Lim)
1.1 million (Potential)

Mongkannwe (Shan East)
Fe -39- 66 % (He.Lim)
21.5 million (Potential)

Pinpeg (Shan South)
Fe -56.4 %(He.Lim)
80 million (Probable)

Maputae Island
(Tanintharyl)
Fe -42 %(He.Lim.Mag)
1 million (Probable)

INDEX

Iron

Iron Occurrences   = 393
Potential   = 495 million tons

Figure 5.8: Distribution of Iron Deposits

DISTRIBUTION OF MANGANESE DEPOSITS

Kyaukpadaung (Mandalay)
Mn - 51.%
0.0115 million (Possible)

Monpyin (Shan South)
Mn - 38.75%
0.096 million (Probable)

Tar Pin (Shan East)
Mn - 6.6%
0.65 million (Possible)

Wansaw Wanpaing (Shan East)
Mn - 12.53%
4.95 million (Possible)

Areye (Shan East)
Mn - 25%
1 million (Possible)

Wansalot (Shan East)
Mn - 14%
0.135 million (Possible)

Powel Island(Tanintharyi)
Mn - 27%
2.8 million (Probable)

Manganese Occurrences= 52

Potential = 11 million tons

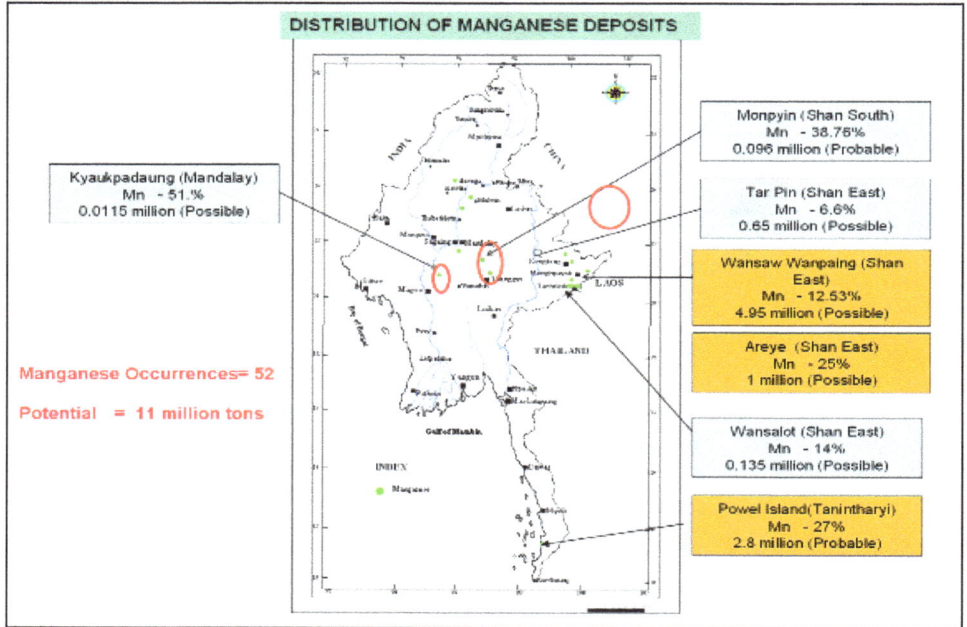

Figure 5.9: Distribution of Manganese Deposits

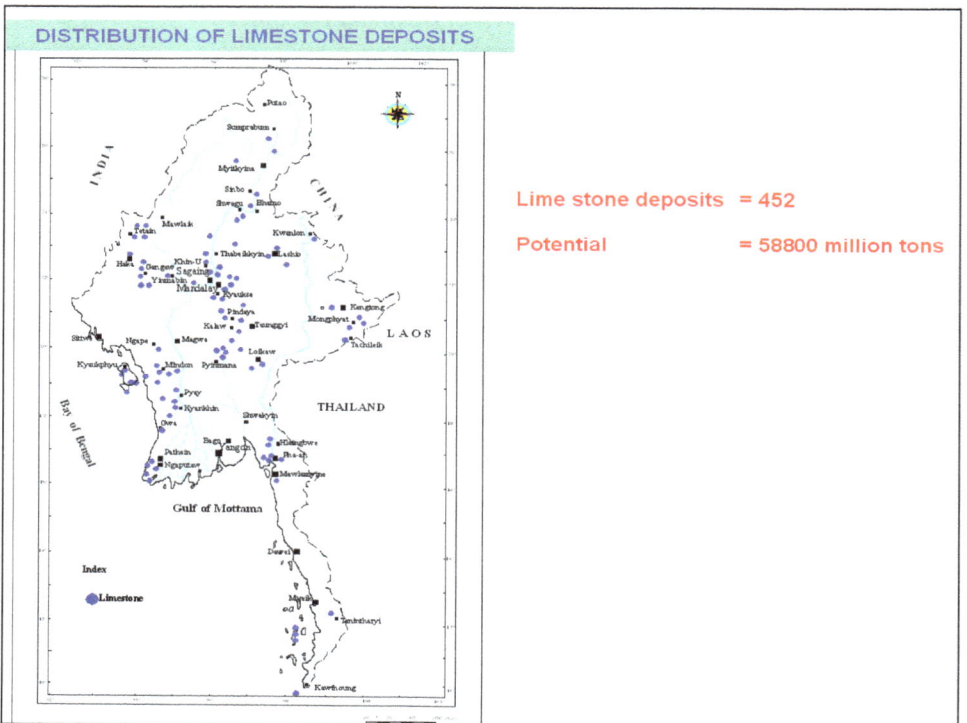

DISTRIBUTION OF LIMESTONE DEPOSITS

Lime stone deposits = 452

Potential = 58800 million tons

Figure 5.10: Distribution of Limestone Deposits

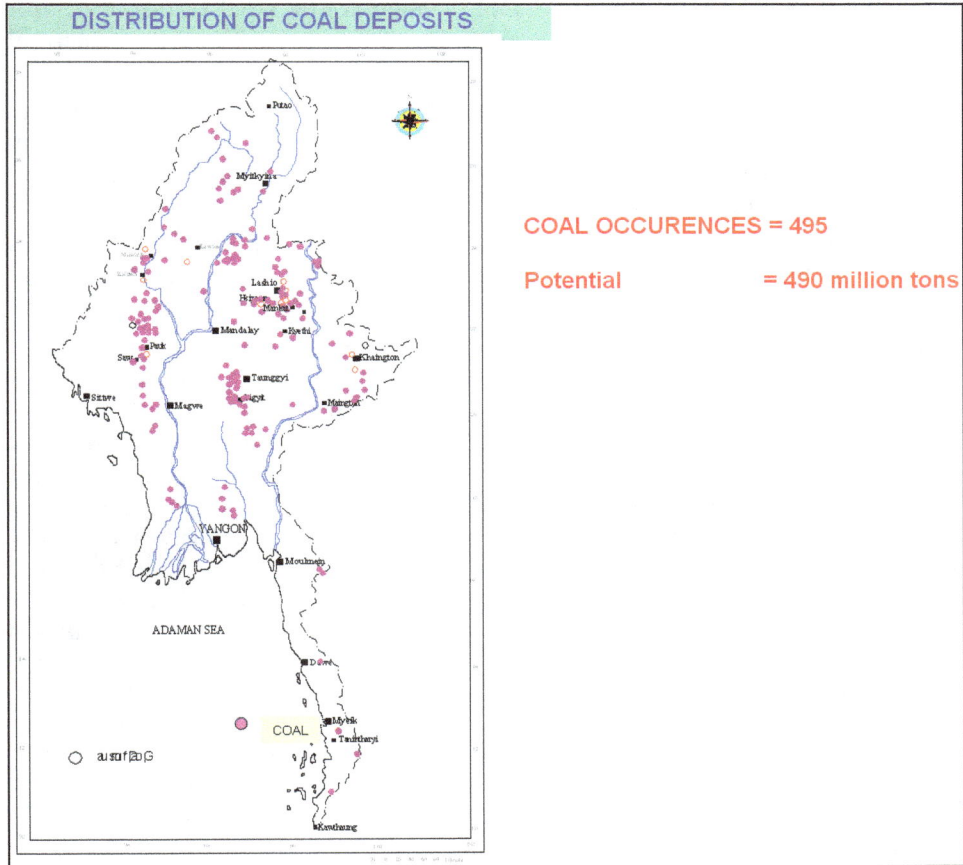

Figure 5.11: Distribution of Coal Deposits

Current Mining Activities in Myanmar

1. Mining Enterprise (ME 1) is to undertake production and marketing of lead, zinc, silver, copper, iron, antimony, nickel, chromite, ores.

2. Mining Enterprise (ME 2) is responsible for production and marketing of tin, tungsten and gold ores.

3. Mining Enterprise (ME 3) is responsible for productions and supply of industrial raw minerals such as barites, gypsum, limestone, dolomite, clay and other industrial minerals, decorative stone and coals.

4. Myanmar Gem Enterprise (MGE) is responsible for mining and marketing of various precious gemstones and jade.

5. Myanmar Pearl Enterprise (MPE) handles breeding and cultivating of mothers of pearl, and production of pearl.

6. Myanmar Salt and Marine Chemical Enterprise (MSMCE) is responsible for production and marketing of common salt, esporn salt, marine chemical and soda ash.

7. Department of Mines is responsible for administration of mineral policy, regulation with ASEAN affairs, planning mineral legislation, mine inspection and safety, issuing mining licenses and environmental conservation in mining.

8. Department of Geological Survey and Mineral Exploration (D.G.S.E) is responsible for country wide geological mapping, mineral prospecting and mineral exploration.

Conclusions

Myanmar is rich in mineral resources. The minerals of potential importance are copper, gold, lead, zinc, silver, antimony, iron, chromium and nickel. The nation's mineral resources remain relatively under-explored. While 86 per cent of the country has been geologically mapped and exploration has not been particularly, carried out in a major way as a result of poor equipment and a lack of secure access to parts of the country. For these reasons, the exact magnitude of mineral reserves in Myanmar remains undetermined. The government controls all mineral exploration, extraction, regulation, and planning through the Ministry of Mines. The ministry of mines is prepared to offer new areas or deposits for new projects and ready to provide raw materials and existing facilities as its participation in the joint ventures. At present, the open-door economic policy of the government has attracted increased number of buyers from abroad. The Extractive Industries Transparency Initiative (EITI) leading authority was formed by a presidential order on December 14 in 2012. As a consequence, Myanmar has applied for EITI membership and now been accepted as a Candidate Country by International EITI Board.

References

1. Economic and Social Commission for Asia and Pacific Atlas of Mineral Resources of The Escap Region Volume 12, Geology and Mineral Resources of Myanmar, United Nations, New York, 1996.

Chapter 6

Industrial Gap and the Mineral Industry in Nigeria

O. Adewole and H.D. Ibrahim

Raw Materials Research and Development Council,
Federal Ministry of Science and Technology,
Abuja, Nigeria.
E-mail: dewolex01@yahoo.co.uk, ceo@rmrdc.gov.com

ABSTRACT

This paper presents a critical review on the potential status and challenges of the mineral sector in. It also presents the way forward to promote the mineral industry in the country. The importance of minerals cannot be overemphasized in view of the fact that it is one of the two major sectors from which all human needs are satisfied, be it a necessity, luxury or even trivialities. It is an established fact that Nigeria is endowed with abundant mineral resources, which if properly managed and exploited, has the potential of becoming the springboard of our non-oil economic growth. These resources range from metallic through, industrial, energy to precious and construction mineral deposits. Worthy of note is the fact that there is no state in the country without the presence of at least one type of solid mineral deposit. Despite the huge resources, Nigeria is ranked 153 in the United Nations Development Index in 2013. The sector, until recently, has not made significant contribution to the national economy. It accounted for only 3 per cent in 2007 and 11 per cent in 2013. The domestic mining industry is underdeveloped. Most of the minerals produced locally are exported as primary raw materials with little or no value addition, thus leading to Nigeria having to import mineral products at exorbitant prices. The mineral industry in Nigeria is characterized by dominance of artisanal mining activities and its attendant shortcomings such as causal and haphazard operations, inefficient mining practices, limited technical and financial capabilities, use of rudimentary and obsolete technologies and human capital. These result in low ore recoveries as well as huge loss of foreign exchange. Efforts undertaken by the government and private investors to transform the economy over the years, including establishment and adoption of institutional, infrastructural, policy and legal frameworks are also discussed.

The mineral sector development has been faced with difficult challenges such as poor infrastructure, difficulties in accessing credit facilities for investment; poor market potential, weak institutional and business supporting framework, poor quality human resource, are highlighted. Others factors hindering the development are inadequate legal, regulatory and judiciary system, lack of appropriate technology for processing of raw materials, apathy towards investment in the manufacturing sector, non – availability of raw materials on sustainable basis and poor entrepreneurship development. Suggested solutions to deal with these challenges are to augment funding for Research and Innovation activities, Investment Promotion through Public–Private Partnership (PPP) collaboration, development of small, micro and medium enterprises, reverse engineering establishment of raw materials processing clusters and establishment of centres of excellence for raw materials R&D.

Keywords: Nigeria, Mining, Industrial gap.

Introduction

Nigeria is the single largest geographical unit in West Africa. It is Africa's most populous country and the 9th most populous country in the world. It has an estimated population of about 160 million people. It occupies a land area of 923,768 square kilometers and lies entirely within the tropics with two main vegetation zones; the rain forest and Savannah zones, reflecting the amount of rainfall and its spatial distribution. The large population is basis for domestic demand for mineral resources, especially industrial minerals (scribd.com (2011).

The fact that Nigeria is endowed with abundant mineral and human resources capable of making her economic giant among the community of nations is an undisputable fact. Apart from rich agricultural land the country is underlain by rock formations playing host to several economic mineral deposits that can be used to transform its economy into one of the leading economies in the world.

Geology of Nigeria

The Basement Complex

About 50 per cent of Nigeria land mass is under laid with the crystalline basement rocks of Precambrian age. These occupy the south western part, the central part of the North and the Eastern boundary of the country. (Gyang, J.D. *et al.*, 2010)

Within the Basement Complex, four major petro-lithological units are distinguishable:

The migmatite-gneiss complexes;

The schist belt (metasediments and meta-volcanics)

The older granites and

The younger granite.

The Migmatite Gneiss Complex

The migmatite gneiss complex is the oldest and the most widespread of the component unit of the basement complex. They are believed to be sedimentary in origin which later underwent prolonged, repeated metamorphism; and now occur as quartzites, marble and other calcareous and relics of highly altered clayey sediments and igneous rocks.It comprises two major types of gneisses: the biotite gneiss and the banded gneiss. The biotitic gneisses are normally fine-grained in nature with strong foliation caused by the parallel arrangement of alternating dark and light minerals while the banded gneisses show alternating light-coloured and dark bands and exhibit intricate folding of their bands.

The Schist Belt

The schist belt (metasediments and meta-volcanics) occur as belts of roughly north-south trending, slightly metamorphosed ancient Pre-Cambrian sedimentary and volcanic rocks known as the schist belt. The major rock types are quartz-biotite-muscovite schist which laterally change into coarse-grained feldspar-bearing micaceous schists. Occurrences of schists with graphite, phyllites and chlorite as well as ferruginous quartzites and talc schists are common. The schist belt play host to most of the gold deposits in Nigeria. Rich gold deposits are found around Maru, Anka, and Zuru in the northwest and also at llesha district in southwestern Nigeria. (onlinenigeria.com)

Older Granites

Older granites occur as large circular masses within the schists and the older migmatite gneiss complexes. They vary widely in composition.

The Younger Granite Complexes

The younger granite complexes are found mainly on the Jos Plateau area of Nigeria forming a distinctive group of intrusive and volcanic rocks that are bounded by ring dykes or ring faults. They are believed to be of Jurassic age. There is a wide variety of composition in this granite series.

Pegmatites

Distributed widely within these major rocks are the pegmatite bodies. They occur as dykes and veins in and vary in size from few centimetres to several metres wide and extend near a couple of kilometers in lengths. They are rich in feldspar with minor occurrences of muscovite and quartz. They also harbor several economic minerals such as gemstones and precious metals.

Sedimentary Basins

The remaining one half of the country is covered by Mesozoic and younger sediments which occur mainly in sedimentary basins including:

- ☆ Anambra Basin
- ☆ Benue Trough
- ☆ Benin (Dahomey) Basin

☆ Bida basin

☆ Bornu (Chad) Basin

☆ Niger Delta basin and

☆ Sokoto Basin

These basins are filled with different types of sedimentary rocks including sand stones, limestone, shale, clays which collectively have provided favorable geological setting for hosting of a wide range of economic mineral deposits; these deposits constitute a variety of raw materials for diverse industries. These include limestone, gypsum, coals of different grades, phosphate nodules, clays *etc.*

Importance of Minerals

The importance of minerals cannot be overemphasized in view of the significant role played by minerals in human lives. What would our lives be like without mining? Imagine a world without jet planes, railroads, cell phones, radar, and jewelry, skyscrapers, parking garages, missiles, and submarines, without medical care items such as X-rays or surgical tools or without electricity generation and transmission and countless other services. We wouldn't have any of these things without mining and use of minerals. Everything we depend on is either made from minerals or relies on minerals for its production. Minerals use and production is also essential in terms of the livelihoods provided through employment and income generation

Table 6.1: Classification of Mineral Resources

Classes	Sub-Classes	Examples
Metals	Iron and ferroalloy metals	Iron ore, Manganese, chromium, nickel, tungsten, molybdenum, vanadium, cobalt
	Non-ferrous metals	Copper, Lead, zinc, Tin, Bauxite (aluminium)
	Precious metals	Gold silver and platinum Group of metals
	Minor Metals	Antimony, beryllium, lithium, magnesium cadmium, Niobium, tantalite, titanium and zirconium
Non metals	Industrial and manufacturing	Mica, talc barytes, glass sands diatomite and bentonite,
	Ceramic materials	Clay, feldspar, talc, Bauxite,
	Gem stones	Tourmaline, beryl, ruby, sapphire, agate, topaz, aquamarine,
	Energy and petroleum	Coal, bitumen, petroleum, natural gas, and Radioactive minerals (*e.g.* uranium)
	Structural and Building materials	Granite, limestone, gravel, sandstone, marble, gypsum, pigments and fillers
	Chemical minerals	Rock salt, brines, potash (trona), sulphur, calcium and magnesium chloride
	Refractory and Metallurgical minerals	Fluorspar, graphite, silimanite, silica, bauxite, clay
	Abrasives	Industrial diamond, garnet, corundum, emery, silica,

Source: Jensen M.L and Bateman AM.

Classification of Mineral Resources

Minerals are basically classified as metallic and industrial (non-metallic) minerals. The metallic minerals are sub divided into iron and ferroalloy, non-ferrous, precious and minor metals. The non-metallic group is also sub-divided into industrial/manufacturing, energy ceramics, structural/building abrasive and chemical minerals (Table 6.1).

Nigerias Mineral Resources

It is an established fact that Nigeria is endowed with abundant mineral resources, which if properly managed and exploited, has the potential of becoming the springboard of our non-oil economic growth. About 34 different mineral deposits have been identified covering all classes of minerals as listed above. Worthy of note is the fact that there is no state in the country without the presence of at least one type of solid mineral deposit Despite the huge resources, only a few (about 13) including Limestone, barytes, lead/zinc, kaolin, gypsum, gemstones, sand, gravel and granite are economically exploited. The remaining ones, though in demand are untapped or at best tapped in a casual way. Table 6.2 presents a list of mineral resources of Nigeria.

Availability of these minerals opens up opportunities in the following areas:

☆ Exports and use in domestic industries for generation of foreign exchange and internal revenue.

☆ Emergence of new industrial and downstream products.

☆ Greater employment opportunities for Nigerians, particularly in the rural areas where the minerals are found. The multiplier benefits to the citizenry are enormous. In fact, the solid minerals sector can easily be the largest employment sector of the economy, since deposits are abound in virtually every State of the Federation.

☆ Technology transfer and development.

☆ Development of infrastructure, especially in the rural areas roads, hospitals, rail, schools and housing

Overview of Nigeria Mineral Industry

The mineral sector, until recently, has not made significant contribution to the national economy. It accounted for only 3 per cent in 2007 and 11 per cent in 2013. Most of the minerals produced locally are exported raw as primary raw materials with little or no value addition, thus leading to Nigeria having to import mineral products at exorbitant prices that she could produce domestically. It is noteworthy that the volumes of domestic trade deficit and foreign exchange losses resulting from this deficiency are colossal.

The domestic mining industry is underdeveloped. It is dominated by artisanal mining activities which are characterized by the following key features: (Miningfacts. com, *MMSD* Report 2002 and CEEST, 1996),

Table 6.2: Occurrence and Reserve Estimate of some Major Mineral Resources of Nigeria

Sl.No.	Mineral	Locations	Industrial Applications	Reserve Estimates (MT)
1.	Barytes	Benue, Cross Rivers, Adamawa, Yobe, Nasarawa, Enugu, Taraba States	Barytes is used in petroleum well drilling, suitable for glass, paint, and paper making. Production of barium metal Used as alloy in vacuum tubes, deoxidizer for copper, lubricant for anode rotors in X-ray tubes and spark-plug alloys.	7.5 million
2.	Bauxite	Taraba, Adamawa Kebbi, Sokoto, Borno Ekiti, Plateau, Benue and Cross River State	Cables, components, alloys	NA
3.	Bentonite	Yobe, Abia, Anambra, Adamawa, Edo, Imo, Ebonyi, Akwa Ibom, Cross Rivers, Benue, Borno States	As drilling mud	700 million tonnes
4.	Bitumen	Ondo, Ogun, Delta, Edo States	Oil, asphalt and other petroleum products	42 billion
5.	Cassiterite	Plateau, Bauchi, Kano, Cross Rivers, Ekiti, Kaduna, Nasarawa, Taraba States	Used in production of solder, electrical components and, chemicals.	300,000
6.	Clays including: Kaolin, Dickite, Halloysite Montmorillonite, Saponite, Illite and Attapulgite	Different types of clays are widely located all over the country.	Ceramic wares such as, floor and wall tiles, dinner and sanitary wares. Refractory bricks, fillers in paper, rubber and paint industries, Electrical insulator. Foundry sands. absorbent and filtering, insecticide dispersing and Cosmetics	Very large
7.	Columbite	Plateau, Kano, Kaduna, Bauchi, Kogi, Kwara, Nasarawa States	Production of electronic components, mainly tantalum capacitors and as alloys with other metals	N/A
8.	Coal	Benue, Enugu, Nasarawa, Gombe, Edo, Anambra, Abia, and Ondo States	Solid fuel for industrial heating and Power generation, extraction of iron/steel, production synthetic gasoline and in metallurgy and foundry applications	Over 3 billion
9.	Copper Ores (Chalcopyrite, Malachite)	Bauchi, Kano and Nasarawa states	Electrical cables wires and switches.	

Contd...

Table 6.2–Contd...

Sl.No.	Mineral	Locations	Industrial Applications	Reserve Estimates (MT)
10.	Dolomite	Kogi, Oyo, Edo, Nasarawa, Kaduna States and FCT	Floor tiles, cement and chemicals	16,540,000
11.	Feldspar	Ekiti, Kogi, Kwara, Nasarawa, Ogun, Ondo and Bauchi States	Ceramic glazing, glass making and tiles	
12.	Gemstones	Plateau, Bauchi, Yobe, Borno, Ogun, Ondo, Kwara, Kogi and Imo States	Jewelry and decorative objects making	
13.	Gold	Cross Rivers, Edo, Kaduna, Katsina, Kebbi, Niger, Osun, Zamfara States	Jewelry, monetary/store of value, electronic instruments and as an electrolyte in the electroplating industry	N/A
14.	Granite	Plateau, Ondo, Ogun, Bauchi, Borno, Adamawa, Kogi, Cross-Rivers, Oyo and Imo States	Rock slabs for floor and wall tiles, hard core, chippings for concrete products	Very large
15.	Graphite	Adamawa, Kaduna and Taraba states.	Carbon electrodes, plates, brushes and dry cell battery; Pencil, crucible foundry pots, dry lubricant, steel hardener and Electrical Components	
16.	Gypsum	Adamawa, Taraba, Benue, Gombe, Ogun, Imo and Borno States	Cement, plaster of Paris, wall and ceiling boards production	1000,000,000
17.	Iron ore	Kogi, Nasarawa,	Steel and steel products, magnets, high-frequency cores, auto parts, catalyst. and metallurgy	Over 3 billion
18.	Kaolin	Katsina, Plateau, Ogun, Bauchi, Ekiti, Ondo and Anambra States	Ceramic wares production, filler in paper, rubber, soap and paint production	Over 3,000,000,000
19.	Ilmenite	Plateau, Nasarawa and Bauchi States	Used as alloys, for electrodes in arc lights. Titanium is a strong lightweight metal often used in airplanes	
20.	Kyanite	Kaduna Ekiti Oyo States.	Refractory and ceramic products, electrical insulators, abrasives and glass	NA

Contd...

Table 6.2–Contd...

Sl.No.	Mineral	Locations	Industrial Applications	Reserve Estimates (MT)
21.	Lead/Zinc	Nasarawa, Plateau, Taraba, Bauchi, Gombe, Ebonyi, Imo, Kano and Benue States	Lead/acid batteries, gasoline tanks, and solders, seals or bearings. Electronic applications such as, TV tubes, communications equipment, protective coatings, in ballast or weights, ceramics or crystal glass foil or wire, X-ray and gamma radiation shielding and ammunition	10,000,000
22.	Limestone	Enugu, Cross Rivers, Ogun, Edo, Benue, Gombe, Sokoto, Adamawa, Ebonyi, Imo and Yobe States *etc.*	Manufacturing of cement, hydrated lime, soil conditioner, chemicals, extraction of iron ore	1,355,980,000
23.	Manganese ore (Pyrolusite)	Plateau and Nasarawa Bauchi Bayelsa and Cross River and Zamfara states	Essential to iron and steel production	
24.	Marble	Oyo, Edo, Nasarawa, Kogi, Katsina, Niger and FCT	Calcium based chemicals, floor tiles	80,292,000
25.	Mica	Kogi, Kwara, Nasarawa, and Oyo States.	Mainly as electrical insulators	
26.	Monazite	Niger, Plateau Nasarawa and Bauchi states	Solar cells and electricity generation	
27.	Quartz and Silica sand	All over Nigeria	Glass making, electronic components like integrated circuits, transistor and solar cell	
28.	Rutile	Plateau, Nasarawa, Bauchi and Kogi States. Sapphire, ruby, aquamarine, emerald, tourmaline, topaz, garnet, amethyst and zircon.	Production of titanium- a strong lightweight metal alloys used in airplanes and for electrodes in arc lights	
29.	Salt	Nasarawa, Taraba, Enugu, Cross Rivers, Benue, and Ebonyi States	Table and industrial salts, food seasonings and chemicals	1 million tonnes
30.	Talc	Niger, Osun, Kwara, Kogi, Kaduna States and FCT	Ceramic insulators, Fillers in paints and fertilizer industries, Talcum powder base in cosmetic industry	100,000,000
31.	Tantalite	Nasarawa, Kaduna, Kwara, Kogi States	Production of electronic components, mainly tantalum capacitors. As alloys with other metals.	N.A.

Contd...

Table 6.2–Contd...

Sl.No.	Mineral	Locations	Industrial Applications	Reserve Estimates (MT)
32.	Uranium	Cross River, Adamawa, Taraba, Plateau, Bauchi, and Kano States	Power generation,	
33.	Wolframite (Tungsten ore)	Kwara, Kogi and Plateau States	Production of filament in light bulbs, electrical component and equipment	
34.	Zircon	Plateau, Bauchi and Nasarawa states	Used in foundry sands	

Source: RMRDC Industrial studies on Base metal, Iron and steel, and Engineering services sector (5th update, 2006), RMRDC Multidisciplinary committee report of the Techno-Economic Survey on Non-metallic minerals sector (4th update, 2003), and Malomo, 2007.

☆ **Informal Operation:** Many artisanal miners are either individuals or families who typically operate without legal mining titles (concession, claim) or a valid contract with the title holder have and invariably prefer to sell their products in the parallel markets.

☆ **Lack of Education:** Most of the miners are poor and live in rural areas, they are unskilled, possess no formal education, and have limited technical and financial capabilities.

☆ **Lack of capital:** They lack capital and have poor access to markets and support services.

☆ **Inefficient Mining Practices:** Mining is done near surface using labour-intensive, risky and inefficient techniques. It is undertaken haphazardly with mines often abandoned in favour of areas with greater potential. Ore beneficiation employs minimal machinery or obsolete technology. This results in low recoveries and the loss of revenue. However, total recovery is improved by repetitive scavenging and reprocessing;

☆ **Low Productivity and Economic Insecurity:** Artisanal miners exploit marginal or small deposits hence low productivity. Activities are predominantly associated with high value, low volume products (*i.e.* high grade low tonnage deposits with no proven reserves).

☆ **Lack of Safety Measures, Health Care or Environmental Protections:** These Miners have no regard for safety measures, sanitation or health, and have little or no knowledge whatsoever of the environmental hazards caused by their activities. They impact significantly on the environment in many ways, including water pollution through mercury and cyanide pollution, direct dumping of tailings and effluents into rivers, flood threats from improperly constructed tailings dams, river damage in alluvial areas, River siltation, landscape destruction, erosion damage and deforestation.

☆ **Part-time Operation:** Artisanal mining may be practiced seasonally (*e.g.*, to supplement farm incomes) or temporarily in response to high commodity prices.

Nigeria's Mineral Production Statistics

The Table 6.3 shows production level for some of the minerals in Nigeria between 2008-2011. (Kumar R, 2013and Brown T.J, *et al.*). This is expectedly low judging by the size of the resources.

Government Reforms

In an attempt to encourage development of the country's solid mineral resources, the Ministry of Mines and Steel Development was established in 1985. The Ministry formulates policy, provides information on mining potential and production, regulates operations and generates revenue for the government (National Minerals and Metals Policy). The mission of the Ministry is to ensure exploitation of the valuable mineral resources spread across the nation for the creation of wealth for all Nigerians, generating employment, reducing poverty, promoting rural economy and make significant contribution to our Gross Domestic Product while its vision is to transform Nigeria's solid mineral sector into an attractive destination for global capital, attracting Foreign Direct Investment to grow the sector to optimum level.

Table 6.3: Production Statistics for some Minerals in Nigeria

Sl.No.		2008	2009	2010	2011
1.	Primary Aluminium	10,600	12,900	21,200	–
2.	Barytes		19,400	19,000	20,000
3.	Coal	30,000	40,000	44,148	
4.	Feldspar		13,631	10,000	10,000
5.	Gold	2,890	1350	3718	600
6.	Gypsum	38,000	160,000	350	350
7.	Iron Ore	62,000	99,424	50,000	–
8.	Kaolin	–	100,000	100,000	100,000
9.	Lead	6,000	5,200	3,300	–
10.	Columbite/Tantalite	335	300	128	–
11.	Tin	1,800	1,800	1,300	500

Source: World Mineral Production 2006-10; Mineral Production Statistics by Country, 2013.

Some key reforms were carried out to build both institutional and infrastructural capacity within government agencies to provide the necessary geo-scientific support required to provide fiscal and legal/regulatory frameworks that are both equitable and manageable in order to attract new private sector investment into the sector.

The Ministry is structured into the following technical Departments:

☆ Mines Inspectorate (MI)

☆ Mines Environmental Compliance (MEC)

☆ Artisanal and Small Scale Mining (ASM)

Under the supervision of the Ministry, the following agencies/parastatals were created for specific functions/mandates.

1. Ajaokuta Steel Company Limited (ASCL)
2. Council of Nigerian Mining Engineers and Geoscience (COMEG)
3. Metallurgical Training Institute, Onitsha (MTIO)
4. National Iron Ore Mining Company (NIOMCO)
5. National Metallurgical Development Centre, Jos (NMDC)
6. National Steel Raw Materials Exploration Agency (NSRMEA)
7. Nigerian Geological Survey Agency (NGSA)
8. Nigerian Institute of Mining and Geosciences (NIMG)
9. Nigeria Mining Cadastre Office (MCO)
10. The Nigeria Extractive Industry Transparency Initiative (NEITI)

Table 6.4: Functions/Mandates of some Key Departments of MMSD and other Government Agencies/Parastatals

Sl.No.	Department/Parastatals	Mandate/Functions
1.	Mines Inspectorate	Responsible for supervision, management and enforcement mining operations and regulations including detailed exploration, evaluation, mine development and production
2.	Mines Environmental Compliance (MEC)	Establish environmental procedures and requirements as applicable to Mining Operations and monitor as well as enforce their compliance
3.	Artisanal and Small Scale Mining (ASM)	Organizing, supporting and assisting small-scale mining operations through the provision of extension and capacity building services to mining cooperatives on exploration, exploitation, mineral processing entrepreneurial training, environmental management, health and safety issues etc;
4.	Mining Cadastre Office (MCO)	Responsible for administration of mining titles with integrity, in a transparent manner and on a "first come first served' basis".
5.	Nigeria Geological Survey Agency (NGSA)	Acquire and provide geoscience information for economic, social and environmental development of the country;
6.	Nigeria Iron Ore Mining Company (NIOMCO)	To explore, exploit, process, and supply iron ore concentrate to the Steel Company
6.	Nigerian Institute of Mining and Geosciences (NIMG)	To provide trainings to Mining Engineers geoscientists, technicians and operators in solid mineral sector; co-operate and collaborate with the University community in the areas of research and training in the geosciences and also provide quality laboratory services to the public and private mining operators.
7.	Bitumen Project	To explore and develop the tar sands and bitumen deposits.
8.	Raw materials Research and Development Council	To accelerate the rate of industrial development through optimal utilization of locally available raw materials resources as input in industries
9.	Nigeria Extractive Industries Transparency Initiative (NEITI)	To promote due process and transparency in extractive revenues paid to and received by government as well ensure transparency and accountability in the application of extractive revenues.
10.	National Steel Raw Materials Exploration Agency	To carry out the exploration of steel raw materials in all parts of Nigeria and elsewhere for the iron and steel industry
11.	National Geosciences Research Laboratories (NGRL)	To provide quality laboratory services to all stakeholders in mining sector
	Council of Nigerian Mining Engineers and Geoscience (COMEG)	Prescribing and enforcing the minimum standards, of education and experience to be obtained by persons qualified to practice as registered mining engineers, geoscientists and metallurgists; enforcing the code of conduct of registered geoscience professionals as well as regulating and controlling the practice of mining, metallurgy, petroleum engineering and geosciences in all aspects and ramifications;

Contd...

Table 6.4–Contd...

Sl.No.	Department/Parastatals	Mandate/Functions
12.	Ajaokuta Steel Company Limited (ASCL)	To produce quality steel raw materials for the nation's steel rolling mills
13.	Metallurgical Training Institute, Onitsha (MTIO)	To build the capacity of Nigeria's metallurgical Engineers
14.	National Iron Ore Mining Company (NIOMCO)	Mining and processing of iron ore to feed the steel company
15.	National Metallurgical Development Centre, Jos (NMDC)	Carry out carry out intensive studies and research on metallic minerals and their process technology

Source: Culled from National Mineral and Metals policy 2007.

In addition to these agencies, the Federal Government has also established other state-owned organizations under relevant Ministries to carry out specific functions aimed at to further developing the mineral sector. These include Raw Materials Research and Development Council (RMRDC) under the Federal Ministry of Science and Technology and Nigerian Export Promotion Council under the Ministry of Trade and Investment. Functions/mandates of these agencies are listed on Table 6.4.

Other Reforms

Other reforms carried out by the government include:

i. **Mineral and Metals Policy**

Mineral and Metals policy was formulated in 2007 with the following objectives:

☆ Achieve a substantial increase in GDP contribution by the minerals sector;

☆ Generate quality Geoscience data;

☆ Establish transparent licensing regime;

☆ Formalize artisanal and small scale mining operations;

☆ Poverty eradication through ASM operations;

☆ Employment generation;

☆ Wealth creation through value addition;

☆ Increase capacity of mineral based industries;

☆ Attract private investment capital;

ii. **Legal reform** – The mining code was revised to bring it in line with global best practice.

iii. **Characterisation studies** – studies were carried out with a view to understanding of both the informal and organised mining activities in the country.

iv. **Capacity building** – Formalizing of artisanal and small-scale miners into mining cooperatives/quarry associations

v. **Registration of mining Operators**- Over 2,000 applications have been received, more than 300 mining cooperatives and quarry associations registered.

vi. **Establishment of Mineral Buying Centres** - Mineral Buying Centres were established to standardise marketing of mineral commodities.

vii. Improved security of tenure through modern Mining Cadastre

viii. Entry visa regime liberalised, visa can be granted up to 10 years.

These reforms have started yielding some dividends for the National Economy

Available Data

The Nigerian Geological Survey Agency has made geo-scientific data available for investors and other stakeholders in the following areas (Malomo S. 2012):

Data in geophysics include airborne magnetic, airborne radiometric and ground gravity all in digital format.

i. **Airborne Geophysical survey** – Airborne survey of the entire country comprising of three components -radiometric, gravity and magnetic which are useful for the following areas and are all available in digital format:

☆ Mineral and hydrocarbon exploration

☆ Groundwater exploration.

☆ Engineering and land use planning

☆ Pollution monitoring

☆ Mapping and monitoring of potential hazards

ii. **Geochemical data** include stream sediments, soil samples and rock samples each complete with samples and analysis results.

iii. **Geological data in form of maps** on scales of 1:200,000, 1: 250,000 and 1: 100,000. Also available as geological data are well logs, cores and rock samples including ditch cuttings, samples, thin sections and analysis results.

iv. **Remote sensing data** include satellite and radar imagery all in both print and digital format.

RMRDC Efforts in Mineral Sector Development

The key mandate of RMRDC is to promote, support and expedite industrial development and self-reliance through optimal utilization of local raw materials as input for the nation's industries. In order to achieve this vision, the Council has initiated and executed a number of mineral based projects some of which are shown on Table 6.5.

Challenges of the Mineral Industry in Nigeria

The challenges confronting the Nigerian economy and indeed the mineral sector are diverse and enormous. With given endowments of natural and human resources, the Nigeria's immense potentials could be realized if these resources are efficiently harnessed and managed. Some of the problems include:

Over-dependence on Petroleum Resources

Nigeria's economy is sustained by mono-product resource-the petroleum. Petroleum resource plays a significant role in the Nigerian economy. The country is the 6th largest producer of petroleum in the world; it is the 8th largest exporter and has the 10th largest proven reserves. While the revenues made from oil provide the largest source of income for Nigeria, the country has become overly-dependent on its oil sector, whereas other areas of the economy such as agriculture, mining and manufacturing suffer near total neglect.

In spite of the country's vast oil wealth, the majority of Nigerians are poor with over 71 per cent of the population living below the poverty line of less than one dollar a day income and 92 per cent or less than two dollars a day. This is indicative of a

Table 6.5: Some Mineral Based R&D and Projects Funded by the Council

Sl.No.	Titles	Collaborators
1.	Establishment of industrial/pharmaceutical kaolin processing plants in Kankara, Katsina State	Katsina State Government
2.	Establishment of industrial kaolin processing plants in Gwarzo, Kano State	Kano State Government
3.	Establishment of Talc Processing model factory in Kagara, Niger State	Niger State Government
4.	Kaolin and Gypsum processing plant Bauchi	Cooperative Society
5.	Establishment of phosphate processing plant Sokoto and Ogun States	Sokoto and Ogun States Government
6.	Upgrading Indigenous Salt Production/Salt processing, Nasarawa State	Nasarawa State Government
7.	Granite Quarrying, Yanyan, FCT, Abuja.	Yanyan Granite Quarrying, Cooperative society
8.	Development of Ceramic Glaze	FIIRO, Oshodi Lagos
9.	Development of Kick Wheel	Local Fabricator
10.	Production of Technical Briefs on several minerals	In House Publication
11.	Production of school Pencils from Nigeria Graphite deposits	PRODA, Enugu State
12	Salt processing project Nasarawa State	Federal Polytechnic, Nasarawa
13	Beneficiation of tin slag for the production of Niobium-Tantalite ingots	
14	Hydrated Lime Project	

Source: RMRDC Annual Reports.

skewed distribution of Nigeria's wealth and regarded as 'resource curse or Dutch disease'. The country is characterized by low sustainable development indices such as high unemployment rate, high incidence of poverty, poor condition of physical infrastructure, health and educational facilities as well as socio-economic, security and governance challenges.(N J Coppin *et al.*, 2005). *Source: Onwualu (2010).*

The 2013 United Nations Human Development Index ranked Nigeria 153 out of 187 countries. (United Nation 2014)

Infrastructure

Supportive infrastructure is an important pre-requisite for successful industrialization. Primary infrastructures such as road network are poor, inadequate and in state of disrepair, communication networks are still inefficient and inadequate, power supply is erratic, unreliable and costly, portable water supply is insufficient and waste disposal facilities are grossly inadequate. Secondary infrastructures such as manufacturing facilities are in short supply.

Institutional and Business Supporting Framework

Existing technological and business promoting institutions which are expected to offer technical advice and innovation in their areas of expertise to the mining sector are deficient in many aspects. This, to a large extent, retards the growth of the sector. Private R&D firms are virtually non-existent; Links among these industrial R&D institutions and the private mining companies are seemingly weak and are usually ad-hoc in nature. They are normally driven by necessity and often dictated by prevailing circumstances rather than inherent technology strategy and/or deliberate plans. Many of these organizations are under institutional reforms, moving from their reliance of government funding to independent, self-financing and autonomous organizations. The institutional framework also suffers from lack of capacity and skills to service the private sector and slow functioning and decision making which greatly affects the performance of the private sector.

Funding Challenges

Investment in mining sector is usually capital intensive and consequently, fund sourcing is a necessary aspect of the enterprise. However, it is very difficult to access credit for working capital from the financial institutions for mining ventures. Presently, the prime lending rates fluctuate between 16 – 25 per cent in some banks. (cenbank.org) This high lending rate only favours service businesses such as trading and imports rather than productive ventures like mining.

Lack of Appropriate Technology for Mineral Processing

Access to some minerals may be constrained by their mode of occurrence. This may require fairly high technology for effective exploitation. Absence of appropriate technology for mining and processing, the obsolete nature of existing machinery and equipment and high cost of acquisition of new ones as well as lack of necessary spare parts for their maintenance constitute great hindrance to mining and processing of these minerals. Consequently, the efficiency, production capacity and product quality of the industries are adversely affected.

Apathy towards Investment in the Mineral Sector

Due to long gestation period of mineral projects, there is great apathy towards investing in this sector of the economy. Preference is rather given to trading than to real sector. This is because of the get-rich quick attitudes of the people at the expense of enduring and sustainable economic ventures.

Cost of Generating Geoscientific Data

Cost of generating geoscientific data is high and the chance of making discovery is low. This resulted in non-availability of reliable essential information on these mineral deposits thus discouraging investment in the mineral sector, particularly from foreign investors. Therefore, a lot need to be done to improve on generating quality data with a view to boosting investors' confidence.

a) Marketing Challenges

Most of the minerals mined in the country are sold with little or no value addition and often at low prices. Dealers are also penalized for not meeting the standards. Minerals being global commodities of which no country is in full control of their prices. Price fluctuation is a common phenomenon the impact of which may be severe on operators especially the artisanal miners.

b)   Entrepreneurship Development and Human Resource Quality

Entrepreneurs are the central actor in industrial activities. On average, Nigeria's industrial sector is characterized by a dearth of entrepreneurial cadre exposed to advanced industrial culture and well trained and experienced human resource in the various fields of raw materials development. This has negatively affected the level of activities in the exploration and exploitation of the abundant mineral resources available in the country. Efforts should be made to build the capacity of our human resource.

c)   Legal, Regulatory and Judiciary System

The legal framework regulating the mineral sector is not comprehensive and partly outdated. Until recently, the existing regulatory framework has been ineffective in catering the needs of industries in terms of licensing, registration, sales, contractual relations, credit, security, property rights, and dispute settlement. The new environment with many actors calls for a greater capacity on the part of public institutions to regulate various actors. This issue is however, being addressed by the current policy framework

Socio Political challenges

Issues like communal strife, civil unrest, boundary disputes and interference by state governments are some of socio-political problems hindering the development of the mineral sector in Nigeria

Investment Climate

The consequence of all that have been said above is the poor investment climate in the economy that has rendered the economy uncompetitive. In the absence of adequate infrastructure (power, roads, water, *etc.*) the cost of doing business in the country remains high.

The Way Forward

In order to ensure sustainable mining sector development, concerted efforts should be made to address the numerous problems bedeviling the sector. These include:

Diversification from Mono Product Economy

Mineral resources are the foundation upon which an industrialised economy is built, and industrialization is essential if the Nigeria is to reduce over-dependence on the oil industry – an industry which, despite the revenue it generates, provides employment for just 6 per cent of the country's labour force. The government recognizes that over-dependence on oil also leaves the economy vulnerable to international oil politics and fluctuations in oil prices. The current efforts aimed at diversification of the economy should therefore be sustained by ensuring consistent policy and implementation of necessary concessions that will further attract investment into the industry by both old and new investors.

Establishment of Credible Geoscience Database

Quality and credible geoscience database is essential for effective development of the mineral sector. The geoscience data from private companies and individuals should be well integrated with national geoscience data base. These synergies will no doubt:

☆ Boost exploration investment by allowing industry to identify areas of favourable mineral potential;

☆ Increase exploration efficiency *i.e.* companies do not duplicate common information or spend money on non-prospective ground

☆ Increase exploration effectiveness by providing key information inputs to risk based decision-making thus reducing exploration costs and risks.

☆ Improve returns on private investment and increase revenues accruing to government as taxes and royalties.

Increased Funding for Research and Innovation

Without adequate funding, R&D capacity building will remain weak (Akinwale Y *et al.,* 2012). Because of the capital intensive nature and risk factors involved, sourcing of funds for mineral project development and processing is a difficult proposition. It follows logically that to strengthen R&D capacity in our research institutes and tertiary institutions (thus making them centres of excellence), adequate funds must be provided to acquire vital tools and equipment and to continually upgrade the skills of R&D personnel. Research funds could be increased through Public–Private Partnership (PPP), increased statutory allocations (UN recommends at least 1 per cent of GDP) and foreign donor agencies. A special fund should be created to tackle the issue of improving competitiveness in our local industries. Such funds can be applied to the acquisition and use of requisite engineering skills and technology for raw materials development. The proceeds of the levy on imports could also be invested in the building of Mineral Technology Institute dedicated to research and capacity building for the researchers in the mineral industry.

Public – Private Partnership (PPP)

There is a need to encourage PPP so as to attract more investment into the mining sector. Greater investment should be made in the development of basic infrastructure under a PPP intervention. Financial institutions in Nigeria should encourage such partnerships by making it easier for them to access soft loan facilities for the establishment and running of the industries. Investors are enjoined to take advantage of the current reforms in the mineral industry and respond favourably to the enabling environment created for the sector by progressive policies of government, and translate the various concessions and reliefs into increased efficiency and increased volumes.to ensure near 100 per cent capacity utilization as soon as possible

Development of Small and Medium Enterprises

For SME's to grow and perform their role as the key drivers of the economy on a continuous basis they must have an access to relevant technology and funds. This is because the globalization of the economy is making it more important to be able to deliver products to customers at a short notice and respond to design challenges in a quick and flexible manner. This situation requires the progressive mastery of new and more sophisticated technologies. Government should, through the relevant agencies, encourage entrepreneurs in the areas of training, credit facilities, acquisition of relevant technologies *etc.*

Reverse Engineering Technology

There is a need for technology adaptation and development in order to effectively develop the capacity of local industries in the country. Countries like India, China, and Japan have gone far ahead in process technology. Our engineers and fabricators can copy such technologies and adapt them for raw materials processing and industrial use. It involves the development of requisite equipment needed for teaching and research in the various engineering fields. This calls for a strong knowledge base on the nature of the various raw materials, their process technology and industrial utilization for production of commercially viable products.

Establishment of Raw Materials Processing Clusters

One way to ensure that raw materials R&D results are translated to industry is to establish raw materials processing clusters. RMRDC recently produced a blue print for this process. The vision is to see the emergence of at least one raw materials (mineral inclusive) processing cluster in every local government in Nigeria. This can exist within industrial parks. In each cluster there will be ten or more processing industries who share common facilities such as power, water, roads, markets, communication, skill acquisition centre, engineering and maintenance centre, finance centre, *etc.* The cluster approach is a PPP arrangement involving RMRDC, State Government, Local Governments, higher institutions and private sector. This is already being driven through the platform of Pan African Competitiveness Forum (PACF) Nigerian Chapter with Secretariat at RMRDC.

Capacity Building

Capacity building to ensure availability of well-trained high level manpower in the mineral sector is essential to improve their performance, increase their productivity and reduce cost of operation.

Conclusions

Nigeria is well endowed with large deposits of mineral resources capable of making her an economic giant but she has not been able to achieve this due to the fact she has not been able to effectively harness these mineral resources.

Necessary strategies and policies are needed to be put in place to speed up activities in this sector. The best way to achieve this is to vigorously pursue research and development progammes in the development of these mineral raw material resources.

Establishment of Raw materials Research and Development Council or similar agencies to coordinate these activities will bring about expected results.

Adequate funding of Research and Development activities and exploration programmes as well as the establishment of Raw Materials Information System and centres of excellence in raw materials development in various countries will accelerate the pace of industrialization.

The challenges of infrastructure need to be properly addressed because supportive infrastructure is an important pre-requisite for successful industrialization.

Conducive regulatory environment for private sector development must be created as a matter of urgency.

Ministry of Mines and Steel Development should effectively enforce adherence to a well-structured solid minerals reporting system, which can be readily computerized so as to ensure credible data base for the sector.

The authorities should increasingly check the activities of smugglers, illegal miners and legitimate operators who submit false returns, and prosecute culprits accordingly. This will not only instill sanity into the sector, boost Government revenue, but also improve the quality of mining and quarrying statistics

Final Note

From the foregoing, it could be seen that huge investment opportunities exist in the solid mineral sector in Nigeria. Nigeria's situation in this sector could be described with the expression – "the harvest is plenty but there are a few labourers

References

1. Akinwale Y, Ogundari I, Olaopa O and Siyanbola W. (2012) Global best practices for R&D funding: Lessons for Nigeria- Interdisciplinary Journal Of Contemporary Research In Business Vol 4, No 2

2. *Breaking New Ground:* Mining, Minerals, and Sustainable Development : the Report of the *MMSD* Project, Volume 1. Front Cover. IIED, 2002 - Business.

3. Brown T.J, Walters A.S, Idoine N. E., Shaw R.A, Wrighton C. Bide E, T.,- World Mineral Production 2006-10 British Geological Survey

4. cenbank.org: http://www.cenbank.org/rates/mnymktind.asp?year=2009-2-13

5. CEEST, (1996)- Environmental Impacts of Small Scale Mining (CEEST, 1996, 62 p.)

6. Coppin N J and Armstrong W. (2005): Sustainable Management of Mineral Resources Project

7. Sectoral Environmental and Social Assessment -Final Report **FINAL REPOR**

8. scribd.com (2011): Economy of Nigeria www.scribd.com/mobile doc 53418924

9. Gyang J.D, Nanle N and Chollom S.G (2010): An Overview of Mineral Resources Development in Nigeria: Problems and prospects. Continental Journal of Sustainable Development 1: 23 - 31, 2010 http://www.wiloludjournal.com

10. Nigerian Geological Survey Agency (NGSA): Geological Map of Nigeria

11. Jennifer Abraham (2013): Harnessing Nigeria's solid mineral resources in Punch Edition, November 14, 2013

12. Jubilee System Ltd (2014): Massive Opportunities in Nigeria's Solid Mineral Sector- A and E Law Publication Powered by Jubilee System Ltd

13. Malomo, S. (2007). Nigeria Mineral Resources: Paper Presented at the International Workshop on sustainable development of Nigeria's Mineral Potentials.Held at the International Conference Centre, Abuja, Nigeria, 22nd –29th November, 2007

14. Malomo S. (2012) Mineral Resources Exploration in Nigeria: Challenges and Prospects for Investment Delivered at the 3rd RMRDC International Conference on Natural Resources Development and Utilization; 24-26, April 2012

15. Mead L J and Alan M B: Economic Mineral Deposits; 3rd Edition Revised Printing

16. Miningfacts.com: Artisanal and Small-scale Mining (ASM) Overview

17. National Minerals and Metals Policy: Publication of Federal Ministry of Mines and Steel Development 2007

18. OnlineNigeria.com -Nigeria - Basement Complex - www.onlinenigeria.com/geology/?blurb=50

19. RMRDC Industrial studies on Base metal, Iron and steel, and Engineering services sector (5th update, 2006),

20. RMRDC Multidisciplinary committee report of the Techno-Economic Survey on Non-metallic minerals sector (4th update, 2003),

21. Rajesh Kumar (2013) Mineral Production Statistics by Country, 2013

22. Report of the Vision 2020 National Technical Working Group on Minerals and Metals Development

23. scribd.com (2011): Economy of Nigeria www.scribd.com/mobile doc 53418924

24. UNIDO (2009) industrialization Strategies and policies focus of panel discussion in NY on African Industrialization Day (UNIDO 2009)

25. United Nation (2014): List of countries by Human Development Index.

Chapter 7

The Indonesian Centre for Artisanal Mining (INCAM): Helping the Nation to Phase Out of Mercury from ASGM

Y. Yudi. Prabangkara[1], Abdul Haris[1],
D. Krisnayanti[2] and Chris Anderson[3]

[1]*Centre for Mineral Resources Technology,*
Agency for the Assessment and Application of Technology
BPPT 2nd Bldg, 12th floor, Jl. MH. Thamrin 8,
Jakarta Pusat 10340 Indonesia
E-mail: yudiprabangkara@yahoo.com
[2]*Department of Agriculture, The University of Mataram,*
Jl. Pendidikan No. 37 Mataram 83125, Indonesia
[3]*Institute of Agriculture and Environment,*
Massey University, Private Bag 11 222 Palmerston North, 4442, New Zealand

ABSTRACT

Indonesia is one of the many countries where the complex geological processes are at work. It has four major tectonic plates where magma activity takes place along the plate rim, well known as "Ring of Fire". It is no wonder that Indonesia is the host of more than 240 active volcanoes, which are the key to mineral wealth in the country, and made it the top three among the most prospective mineral resource countries in the world (Fraser Institute, 2013).

Currently, The Government of Indonesia has released 39 contracts of work (CoW) for mineral and metals, 76 CCoW for coal and more than 6000 mining licenses released by local government, mostly for large scale mining operations. However, under the new Mining of Mineral and Coal Law No. 4/200, the Indonesian government has made the provision for the artisanal and small-scale mining in the Indonesian Law.

On the ground, historical records suggest that artisanal gold mining has been practiced throughout Indonesia for hundreds if not thousands of years. However, the scale of operation has steadily increased since 2000. In all areas, mercury amalgamation has traditionally been the preferred technology for gold recovery. Mercury has historically been cheap, readily available, and gold recovery using mercury is very simple. A range of NGOs continue to focus on mercury reduction as a key technological intervention. Simple retorts to recover mercury from amalgam balls have been introduced to the artisanal gold-mining circuit and these reduce the release of toxic mercury into the environment. The success of this intervention highlights how simple technology can be effectively used for environmental protection.

However today, mercury is gradually being replaced by cyanidation as the primary gold recovery technology and the widespread use of cyanide is creating new environmental challenges. Cyanide will dissolve mercury and other heavy metals in the gold ore being processed, and is discharged into the environment at the end of the mining cycle. Mercury complexed with cyanide will methylate in rice paddies. The resulting methyl-mercury compound can be accumulated by rice plants exposing the population of mining areas to unacceptable environmental risk.

The sustainability of artisanal gold mining in Indonesia is threatened by poor mining infrastructure, inefficient processing technology, and an absence of waste management. The Indonesian Centre for Artisanal Mining (INCAM) has been established to respond to the explicit need for improvement in each of these areas. INCAM's mission statement is 'to contribute with education and technology to the transformation of artisanal miners into responsible small-scale miners while striving, through a collaborative effort, to help them reduce poverty and improve quality of life for workers and affected communities'. This mission statement mirrors that of the International Training Centre for Artisanal Miners (ITCAM) currently being established in Ecuador.

Introduction to Artisanal and Small-Scale Gold Mining (ASGM) in Indonesia

Indonesia has a rich endowment of mineral resources due to the geographical placement of the archipelago along the Pacific Rim of Fire. Volcanic activity has created mineral deposits containing significant quantities of tin, coal, copper, nickel and gold (Mbendi, 2012). In 2010 Indonesia ranked 7th in terms of global gold production (127 tones) (Widjajanto and Arif, 2011) with capacity for an increase in this output. Many studies have concurred that the amount of the gold produced ASGM resembles the above number although the exact amount is not known.

The ASGM has been practiced in Indonesia for hundreds of years. The steadily increasing gold price has attracted more miners to play an active role in this sector over in the last few decades. As a consequence, hundreds of thousands of miners have now been operating throughout Indonesia.

In term of the legality of their operation, most (if not all) of them are operating without a valid paperwork, making it difficult to control by the government. In addition, they are mobile in nature, where their appearance and disappearance are very much controlled by the gold deposit. Therefore, tracing and and controlling them even requires more laborious and persistent efforts.

Large scale mining companies have usually been licensed to explore and exploit the mines in their respective mining areas. During an exploration phase, those companies usually utilize the services of local people during mapping and drilling. These locals usually then work in the "ex drilling" area even before the large mining company make a financial and operational judgment. This situation has caused several "clash" between large-scale companies and the Artisanal and Small Scale Mining (ASMs).

ASMs also operate in the conservation forests, making it vulnerable to deforestation and environmental degradation as groups of people will usually reside and commute. On the other hand, they also operate in build up areas, where miners make shafts between houses and process and burn the gold in the buildup area. In this area, environmental pollution becomes a matter of concern having social.

The types of gold deposits mined by ASMs also vary from placer to primary deposits. On placer deposits, mining is usually undertaken using hydraulic mining technique, where high-pressure water is applied on loose materials using a pump. Another pump is usually used to suck collected materials and hosed them onto a sluice box. The materials remained in the sluice is then transferred and subject to panning to get gold particles. Miners sometime add drops of mercury during the process to increase the gold recovery. Depending on the form of the gold, they may sell the gold sand directly, or bring the amalgam to the gold shop, where the gold shop will burn the amalgam in their shop and then weigh the gold.

For a primary deposit, the process of mining the ores for the extraction of the gold is even more complicated. Primary gold deposits usually found tens of meters underground, so that miners are required to build a vertical shaft to reach the deposit and then dig the tunnel following the deposit.

The shaft is usually 1 x 1 meters wide, just enough to slip the miners and ores in and out. The shaft is mostly supported by timber to avoid unnecessary rock fall as well as to provide supporting steps for miners to get into the tunnel underneath. On deep shafts, life support is usually provided by blowing fresh air from a blower.

Figure 7.1: Mining Gold from a Placer Deposit Practiced in Sulawesi, Indonesia

Figure 7.2: Sluice Boxes Used to Separate Heavy Material from the Light Ones

Figure 7.3: Amalgamation Process using Trommels is a Common Practice in Indonesia

Figure 7.4: Burning of Amalgam in the Open Air

However, in general, safety is usually neglected, resulting in a number of casualty from accidents like rockfall, and suffocation.

The ores from the tunnel are then crushed using hammer to form smaller ore particles. The ores are then transferred into trommels, mixed with water and mercury. The cocktail undergoes a milling process for several hours in the trommels during which the ores are milled and gold particles are released; react with mercury to form an amalgam. After this process, the amalgam is harvested while mercury-containing tailing is collected for sale and co-processed using cyanide. The amalgam is usually burned in a open air to get the gold.

As the use of mercury in the ASM sector is widespread in Indonesia. It is estimated that the use of mercury in the sector is around 100 – 200 tones per annum, which is in contrast to the registered import of mercury of merely 0.5 tones per annum.

Although the ASGM has been viewed as negative for many, the ASGM sector, in reality bring about positive economic boost to locals, and if this is well regulated, will increase the national income from the gold export.

Based on the above, perspective of a local government (personal communication), a district where ASM has flourished in recent years, ASGM has led to:

- ☆ increased employment
- ☆ reduced criminality
- ☆ increased community income
- ☆ generated alternative livelihood during dry season
- ☆ increased participation in schooling
- ☆ increased purchasing power of local community

☆ reduced conflict in community

☆ reduced temptation to become migrant workers overseas.

From the above it can be seen that positive impacts of ASGMs has to be celebrated and welcomed. At the same time, efforts have to be made to ensure that negative impacts of the ASM can be minimized and mitigated.

Mining Sector – A New Era

The government of Indonesia has rejuvenated the mining sector by introducing legislation which replaces the old mining law No. 11/1967 with a new one, Mineral and Coal Law No. 4/2009. There is a new dimension in the new mining law, by accommodating People Mining, a term which was not accommodated in the 1967 law. These laws not only make provision for the people engaged in mining activity, but also distribute the permits to local government. This law also imposes ban on the export of raw mineral, which in turn, boosts the establishment of mineral processing and refining industries.

In the context of ASGMs, the new mining law also ensures that artisanal and small scale gold miners can apply for a permit, as long as they fulfills the requirements and operate in areas designated for people mining. Therefore, the formalization of the artisanal miners into the national economic mainstream can be undertaken legally.

Secondly, the government of Indonesia has signed a legally binding document Minamata Convention on mercury elimination in October 2013. The government of Indonesia and other signatories have agreed to reduce the use of mercury in all sectors by 2018.

In order to make it operational, the Government of Indonesia has released a National Action Plan on the mercury elimination. The government, through the ministry of Energy and Mineral Resources, has also formulated a National Action Plan (NAP) on mercury elimination from mining sectors.

The NAP has three major components:

1. Legal framework and institutional strengthening,
2. Research and development,
3. Increased awareness and communication.

The second Component, Research and Development, is the area where The Agency for the Assessment and Application of Technology (BPPT) is a key player. The function of this component is to provide a mercury-free-technology, which can be used and adopted by miners without having to invest too many new equipment and can be operated by miners. The challenge would be to provide the miners with a new (or even old) technology that produce more gold, without sacrificing too much their lifestyle at the moment.

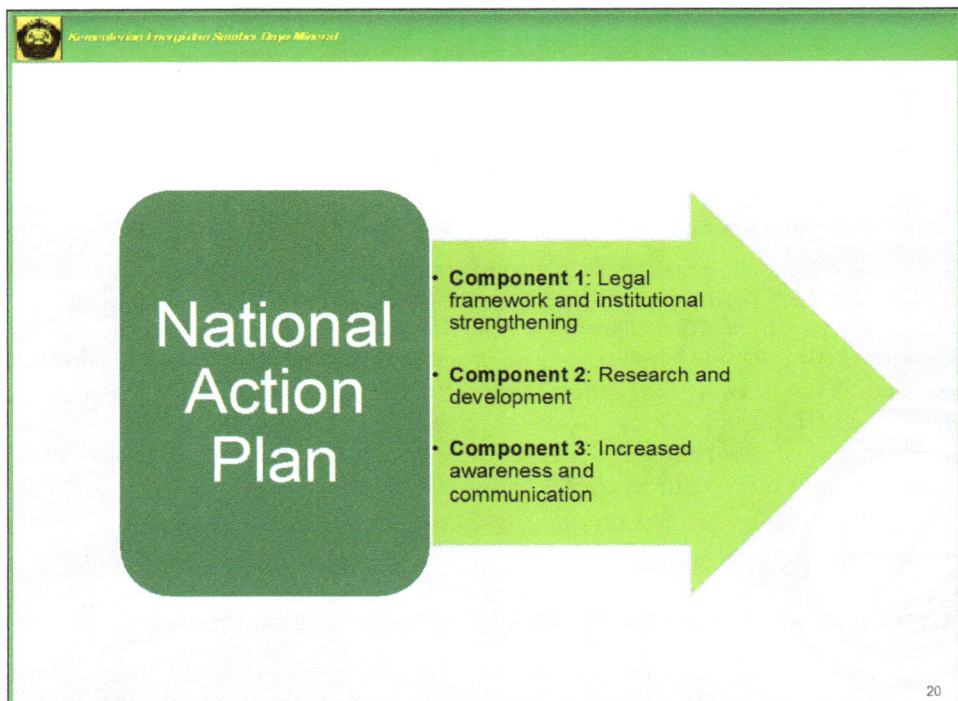

Figure 7.5: National Action Plan for Mercury Elimination from ASGM
(*Source*: **Kementerian ESDM, 2014**)

One of many important components in the national action plan is how to effectively communicate the above components to the ASGM community *i.e.,* ***Awareness and Communication***. Effective communication will not only help to phase out the use of mercury, but also to transform the ASGM into a more responsible Small scale mining (SSM).

The first program in the, third component is to increase the *technical capability* of the miners to adopt new technology. This will be undertaken through socialization and training on how to safe handling of chemicals used in the gold processing, and on the mercury-free technology. Based on the national action plan, this should be undertaken between 2014 – 2018, where BPPT, Ministry of Energy and Mineral Resources, Ministry of Environment and and the Indonesian Institute of Sciences are active stakeholders.

Another important program in the component 3 is an effective communication to miners. The program will include, to create social awareness of the environmental and health risk of the use of mercury as well as non – mercury techniques, using electronic media and associated media, on the job training for communities around ASGM location for alternative livelihood. To have a long term effect, an early childhood education will be important for the sustainability of the program and for ensuring that mercury will not be used by future generations. This can be undertaken by developing curricula on mercury effect in elementary schools. This will be collective efforts of central and local stakeholders.

Creating a Sustainable ASGM in Indonesia

Artisanal gold mining is a livelihood for poor communities around the world. Miners respond to international demand for mineral commodities and exploit resource using technology and techniques available to them. Miners do not choose to be environmentally destructive; the harmful consequences of ASGM are due to unsafe and unregulated mining systems.

We believe that the solution to the problems of ASGM lies in education, and there is precedent for large scale empowerment of miners with knowledge and skills. For example, the UN Global Mercury Project educated over 30,000 miners in six pilot countries to adopt more efficient methods to recover gold and generate less pollution (McDaniels and Veiga, 2010). Recently, the University of British Columbia implemented a demonstration gold processing plant in Ecuador to train more than 600 artisanal miners from Peru, Columbia, Ecuador and Bolivia. The result has been a reduction in mercury pollution by up to 63 per cent and provides a catalyst for the establishment of cleaner and more efficient gold processing plants in Columbia (39) and Ecuador (40).

Education is therefore essential to maximise the benefits brought about by artisanal mining and to avoid or mitigate the negative impacts. The technology to solve many of the environmental and social problems related to ASGM exist; what is lacking, in the opinion of many government, non-government and academic commentators, is a desire to effect a positive change (Hylander *et al.*, 2007).

Mechanism for Sustainable Development: Training Centres for Artisanal Miners

The International Training Centre for Artisanal Miners (ITCAM) initiative was conceived by the Department of Mining Engineering at the University of British Columbia in Vancouver, Canada, based on joint review of UBC's successful ASGM education program by the Governments of Ecuador, Peru and Columbia of the success. The purpose of this review was to explore the possibility of establishing formal and sustainable educational institutions in the artisanal gold mining regions of each of these countries. The objective of the global ITCAM initiative is to integrate educational and training curricula with appropriate technology to address global problems related to artisanal mining.

Development of INCAM: The Indonesian Centre for Artisanal Miners

In May 2013, The International Research Centre for the Management of Mining and Degraded Lands (IRC-MEDMIND) hosted a workshop on Artisanal and Small Scale Gold Mining (ASGM) at the University of Mataram. The workshop was convened to draw an action plan for the sustainable future of ASGM in West Nusa Tenggara Province.

Approximately, 100 delegates at the workshop agreed that ASGM was now a permanent activity in WNT Province, and that the economic impacts of the mining were generally positive. However, there was agreement that the environmental and social impacts of mining must be better managed. The workshop developed a clear goal: to establish a Centre of Excellence for ASGM that was to provide training and advanced technology to ensure that ASGM mining communities and the environment are safe, healthy and prosperous. The Centre of Excellence was given the mission statement, from inception, to support the sustainable development of ASGM throughout Indonesia through focusing on mining practices, technology to recover gold from the rock, and the safe management of mining waste.

Key proponents of the ITCAM initiative from the NBK Institute of Mining Engineering, University of British Columbia were present at the IRC-MEDMIND workshop. UBC contributed to the discussion, and proposed that the Centre of Excellence be an Indonesia-targeted roll-out of the ITCAM. This proposition was subsequently endorsed by workshop participants and development of the Indonesian Centre for Artisanal Miners was begun. The Mataram workshop stipulated that the Centre should be led by the Government of Indonesia in collaboration with national and international university, private-sector and NGO experts; have education and technical-training embedded as key responsibilities; and should be supported by international aid programs.

The INCAM mandate is to:

☆ Promote global, national and local awareness of artisanal mining issues

☆ Develop appropriate technological and educational programs to meet local needs

☆ Help artisanal mining communities improve their quality of life

☆ Create opportunities for alternative livelihoods and enterprise in mining communities

☆ Help artisanal miners increase mineral recovery, increasing profit margins

☆ Help artisanal miners reduce the environmental, social, and health impacts of their practice

☆ Assist governmental agencies in understanding technical issues in artisanal mining

☆ Assist mining companies address regulatory problems related to artisanal mining

INCAM seeks to transform Indonesian artisanal miners into socially- and environmentally-responsible small-scale miners.

INCAM will integrate government, industry, university and NGO aspiration to create a sustainable future for ASGM in Indonesia. Through research and training, technology, best practice, efficiency, and environmental management, a sustainable future for artisanal mining will be created. A significant focus for INCAM will be ASGM, but mechanisms to ensure the long-term sustainability of artisanal mining for other commodities, such as nickel, tin, coal and aggregate, will be also be developed.

INCAM initiative will play a significant role supporting the National Implementation Plan team to meet its objective. INCAM has the mandate to identify, develop and implement non-mercury mining technology throughout Indonesia. The Ministry of Environmental has requested that INCAM become a formal and legal part of the National Implementation Plan.

Indonesian Training Centre for Artisanal Miners (InTCAM)

INCAM will have a training centre arms in fields called Indonesian Training Centre for Miners (InTCAM). The training centre division of INCAM will be responsibility for implementing technology options to field locations.

InTCAM will implement an appropriate educational curriculum in partnership with INCAM's international training providers (Massey University, New Zealand; the University of British Columbia, British Columbia, Canada; and the College of the Rockies, British Columbia, Canada). Based on the experience of ITCAM operations in South America, key infrastructure within the training centre will be a mineral characterization lab (MCL) and a mineral pilot processing plant (MPP) capable of processing 5 tons of gold ore per day with purpose built tailings storage facility. MCLs established at ASGM locations in Indonesia will enable miners to quantify the gold (and other commodity) content of ore (grade control). These labs will support technology transfer and training in efficient and environmentally-sustainable mineral processing.

An additional responsibility of InTCAM will be to train local NGOs and local government in the necessary skills and to monitor health of both miners and the environment in artisanal mining areas.

The INCAM project has identified west Sumbawa as the first location for its training centre. This centre will be established on 1.65 hectares of land at Poto Tano, 20 kms from Taliwang township (Figure 7.6) that has been loaned to BPPT by the Government of West Sumbawa for a period of 20 years. The training centre will be run as a partnership between INCAM, the Bupati (regional government) and the mining company PT Indotan which is active in the area. The partnership will establish the training centre as a local enterprise for the duration of the lease period. Therefore, management of the Poto Tano centre will be the responsibility of the regional government. Indotan will build the pilot plant and provide ore from mining operations for use in the training centre.

The training centre model for Poto Tano will be used by InTCAM as a template for training centre development throughout Indonesia; there is demand for training centres in Kalimantan, Sumatra and Java. Each training centre will be established as a partnership between INCAM (technology and curriculum), government (land) and the mining sector (pilot plant and ore). Training centres will educate and empower local miners and communities with safe and efficient mining skills. These miners will be able to approach mining as a profitable business in harmony with the established mining private sector. Association of the regional and district government with the training centre will ensure appropriate permitting and regulation of mining activities.

Figure 7.6: Location of the Proposed Training Centre in West Nusa Tenggara Province

Conclusions

☆ Indonesian is rich in mineral resources, and one of world most prospective countries for mineral resources.

☆ Gold production has been a major export commodity, produced by large scale mining companies. In addition, relatively equal amount of gold has been produced and exported by ASGMs.

☆ While ASGMs have been the single largest contributor to the use of mercury in Indonesia, the sector is undeniably, help the government in increasing the livelihood of the community.

☆ Indonesia has formulated a national action plan for mercury elimination from ASGM sectors. Of the major components presented in the action plan, research and development and increased awareness and communication. BPPT will have a major role in the development and dissemination of mercury-free technology to be used by ASGM sectors, by establishing an Indonesian centre for Artisanal Mining (INCAM).

☆ INCAM seeks to transform Indonesian artisanal miners into socially- and environmentally-responsible small-scale miners.

References

1. Hylander, L.D., Plath, D., Miranda, C.R., Lucke, S., Ohlander, J. and Rivera, A.T.F., 2007. Comparison of different gold recovery methods with regard to pollution control and efficiency. *Clean*, 35: 52-61.

2. Mbendi, 2012. Mining in Indonesia – overview. Accessed from http://www.mbendi.com/indy/ming/as/id/p0005.htm Nov 20, 2012.

3. Kementerian ESDM, 2014. National Action Plan for Mercury Elimination from ASGM. Ministry of Energy and Mineral Resources.

4. Krisnayanti, B.D., A. Haris and C. Anderson. 2014. INCAM – Develepment Plan 2014. CRC MEDMIND and BPPT. 37pp.

5. McDaniels J., Chouinard, R. and Veiga, M.M., 2010. Appraising the Global Mercury Project: an adaptive management approach to combating mercury pollution in small-scale gold mining. *International Journal of Environment and Pollution*, 41: 242-258.

6. Widajatno, D. and Arif, I., 2011. The Indonesian Mineral Mining Sector: Prospects and Challenges. Association of Indonesian Mining Professionals.

7. www.mercurywatch.org, 2014

Chapter 8

Zimbabwe and the Platinum Value Chain

Tony Nyakudarika

Principal Process Engineer,
DRA Projects (Pty) Ltd, Johannesburg, South Africa
E-mail: tony.nyakudarika@DRAglobal.com

ABSTRACT

The world's Platinum Group Metals (PGMs) come from South Africa, Russia, North America and Zimbabwe. The PGMs include iridium, osmium, palladium, platinum, rhodium, and ruthenium. Only South Africa, USA and Zimbabwe extract PGMs as primary minerals, with the other players, namely Russia, Canada and China producing PGMs as by-products from nickel-copper operations. Recycling of PGMs is also a significant activity. Therefore, the current worldwide production capacity of PGMs is inadequate to meet the growing global demand. Zimbabwe has a significant role to play in the supply of PGMs and stands to benefit tremendously if it can participate long way down the value chain.

For Zimbabwe to realise the potential of its PGM resources it is critical to understand the extent of the resource and the nature of the global market, including supply and demand situation. Knowledge of the PGM value chain is a prerequisite to extracting value from this resource. The challenge though is keeping abreast with any advances in processing of PGMs and new developments in the industry. This calls for development of relevant skills and co-operation with established players.

There are opportunities and challenges associated with the drive to develop a sustainable PGM industry for the benefit of the entire population, which include, but not limited to development of downstream industries, skills shortage and limited (or unavailable) infrastructure.

PGMs have a wide use in domestic and industrial applications due to their catalytic properties, wear and tarnish resistance characteristics, resistance to chemical attack, excellent high-temperature characteristics and stable electrical properties. The paper is a discussion document on the PGM industry, with specific reference to the value chain and how it impacts Zimbabwe's economy and well being.

Introduction

Naturally occurring platinum and platinum-rich alloys have been known since a long time. The Spaniards named the metal "platina," or little silver, when they first encountered it in Colombia. They regarded platinum as an unwanted impurity in the silver they were mining. Today PGMs play a vital role at the heart of everyday living. It is estimated that one out of four (20 per cent) goods manufactured worldwide either contains PGMs has PGMs play a key role in their manufacture. How noble these minerals are!

Platinum Group Metals (PGMs) are a group of 6 chemically very similar elements comprising the light platinum metals ruthenium (Ru), rhodium (Rh), palladium (Pd) and the heavy platinum metals osmium (Os), iridium (Ir) and platinum (Pt). PGM deposits are found in two main forms, PGM-dominant deposits and nickel-copper sulphide deposits. Platinum and palladium are soft, ductile and highly resistant to heat and corrosion. All PGMs alloyed with one another or with other metals can act as catalyst. These properties of PGMs made them ideally suited for their wide use in industry and domestic applications.

PGMs are sourced from mine production and by recycling of from autocatalysts, electrical components and jewellery. At least 20 per cent of the PGM requirement is met by recycling. PGM-dominant deposits are associated with sparsely dispersed sulphide mineralisation, PGMs being the main economic component with nickel and copper as less valuable by-products. Examples are the Bushveld Complex in South Africa, the Munni Munni Complex in Australia, the Stillwater Mining Complex in USA and Great Dyke in Zimbabwe.

Nickel-copper sulphide deposits of PGMs occur in association with sulphide-rich ores and constitute as by-products. Examples are Lac des lles in Canada, Jinchuan in China and Pechenga district in Russia. Botswana also produces palladium in along with nickel production.

This paper gives an insight into the PGM industry. It is hoped that by understanding the sources, uses and market dynamics participants can engage in a discussion on implications for developing countries from the viewpoint of a knowledge base, and be guided to formulate best strategies for dealing with the challenges and opportunities posed by this noble resource for the benefit of all.

PGMs Ocurrence

The Zimbabwe Great Dyke is increasingly becoming a significant player on the global PGM map. The Great Dyke spreads over about 550 km in a north-south direction along the heart of Zimbabwe. Figure 8.1 is the locality map for the Great Dyke.

Currently Zimbabwe has 3 major PGM players who are associated with mining houses in South Africa, being Mimosa Mining Company, Zimplats running the Selous Metallurgical Complex and Ngezi operations and Unki Mines. An estimated 10 Million tonnes of ore per annum is treated to produce both PGM concentrate and matte. This is equivalent to 811 000 refined oz. of 4E (Platinum, Palladium,

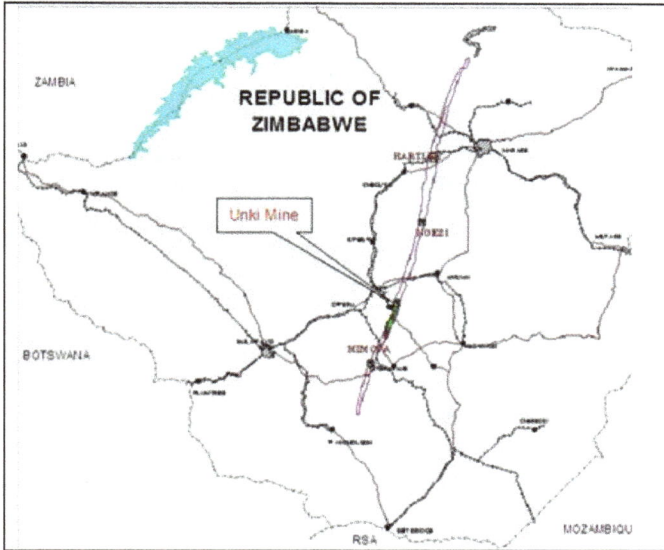

Figure 8.1: Map of Zimbabwe Showing the Great Dyke

Rhodium and Gold) or 340 000 oz. Platinum. It is possible that more players will be involved in exploitation of the Great Dyke PGM resources.

PGM Sources

A discussion on "Zimbabwe and the Platinum Value Chain" would not be complete without understanding the global supply and demand dynamics of PGMs. World resources of PGMs are estimated at 100 million kg (3, 215 million oz tr) in reserves at 66 million kg (2,125 million oz tr) with South Africa as the predominant player. Table 8.1 and Figure 8.2 show distribution of the reserves).

Table 8.1: World PGM Reserves (Zimbabwe included in others)

Country	PGMs Reserves (kg)	Fraction of World (per cent)
USA	900 000	1.4
Canada	310 000	0.5
Colombia	–	–
Russia	1 100 000	1.7
South Africa	62 000 000	95.3
Zimbabwe	–	–
Others	800 000	1.2
World Total	66 110 000	100

Mine Production

Production of PGMs follows the reserve trend with South Africa dominating

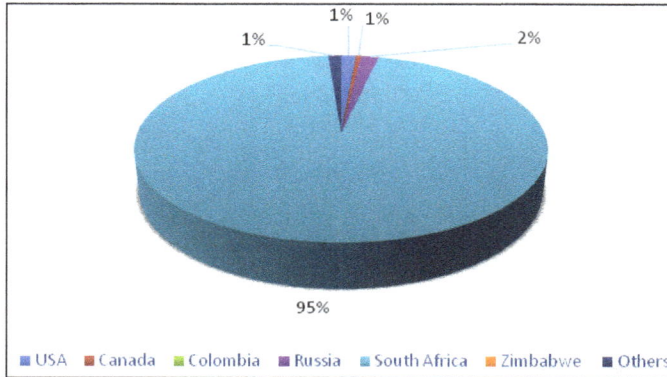

Figure 8.2: World PGM Reserve Distribution

as shown in Tables 8.2 and 8.3, Figures 8.3 and 8.4 for platinum and palladium, respectively.

Table 8.2: Platinum Mine Production

Country	2011 Pt Pdn (kg)	2012 Pt Pdn (kg)	2013 Pt Pdn (kg)
USA	3 700	3 670	3 700
Canada	7 000	7 000	7 000
Russia	25 000	24 600	25 000
South Africa	145 000	133 000	140 000
Zimbabwe	10 600	11 000	12 000
Others	3 730	3 480	4 000
World Total	**195 000**	**183 000**	**192 000**
Recycle	**64 100**	**63 500**	**64 500**

Table 8.3: Palladium Mine Production

Country	2011 Pd Pdn (kg)	2012 Pd Pdn (kg)	2013 Pd Pdn (kg)
USA	12 400	12 200	12 500
Canada	14 000	13 000	13 000
Russia	86 000	82 000	82 000
South Africa	82 000	72 000	82 000
Zimbabwe	8 200	8 900	9 000
Others	12 200	12 000	12 000
World Total	**215 000**	**200 000**	**211 000**
Recycle	**74 200**	**71 200**	**76 500**

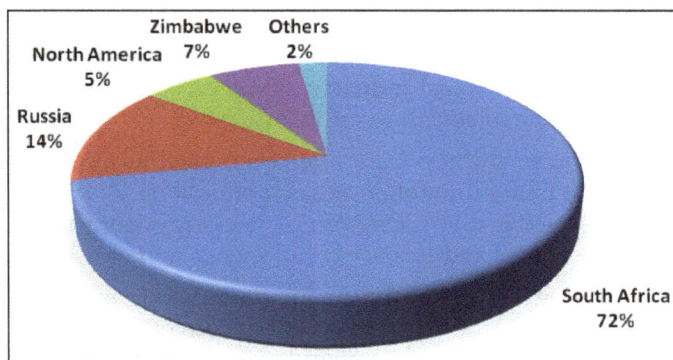

Figure 8.3: World Mine Pt Production 2013

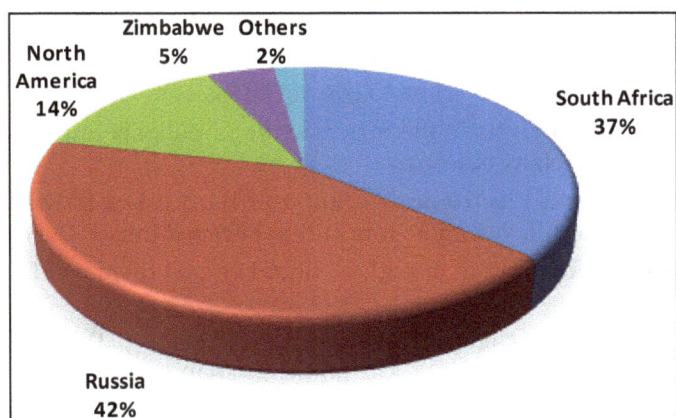

Figure 8.4: World Mine Palladium Production 2013

Understanding The Market – Supply and Demand (the value chain)

PGMs and their alloys have excellent properties. The catalytic properties of the six platinum group metals (PGM) – iridium, osmium, palladium, platinum, rhodium, and ruthenium – are outstanding. Platinum's wear and tarnish resistance characteristics are well suited for making fine jewellery. Other distinctive properties include resistance to chemical attack, excellent high-temperature characteristics, and stable electrical properties.

All these properties have been exploited for industrial applications. Platinum, platinum alloys, and iridium are used as crucible materials for the growth of single crystals, especially oxides. The chemical industry uses a significant amount of either platinum or a platinum-rhodium alloy catalyst in the form of gauze to catalyze the partial oxidation of ammonia to yield nitric oxide, which is the raw material for fertilizers, explosives, and nitric acid.

In recent years, a number of PGMs have become important as catalysts in synthetic organic chemistry. Ruthenium dioxide is used as coatings on dimensionally stable titanium anodes used in the production of chlorine and caustic.

Platinum supported catalysts are used in the refining of crude oil, reforming, and other processes used in the production of high-octane gasoline and aromatic compounds for the petrochemical industry. Since 1979, the automotive industry has emerged as the principal consumer of PGMs.

Palladium, platinum, and rhodium have been used as oxidation catalysts in catalytic converters to treat automobile exhaust emissions. A wide range of PGM alloy compositions is used in low-voltage and low-energy contacts, thick- and thin-film circuits, thermocouples and furnace components, and electrodes.

Fuel cell technology is assuming major dimensions in the application of PGMs. Fuel cells are electrochemical devices that convert the energy of chemical reaction directly into electricity, with heat and water as by-products. Platinum and ruthenium are the key elements of this technology, with research on the use of the other PGMs on-going.

PGMs are used in the following applications.

☆ Autocatalysts – convertors to control emissions

☆ Chemical - catalysts in the manufacture of bulk chemicals such as nitric acid and specialty silicones

☆ Electrical and Electronics – computer hard disks to increase storage capacity, multilayer capacitors and hybridised integrated circuits

☆ Glass – production of fibreglass, liquid crystal displays (LCDs) and flat panel displays

☆ Investment – Exchange Traded Funds (ETFs) (*e.g.* Tokyo Commodities Exchange and New York Mercantile Exchange and most recently SA's ABSA ETF

☆ Jewellery

☆ Medical and biomedical – dental restorative materials, implants

☆ Petroleum – refining

☆ Fuel Cells

Table 8.4: Platinum Demand by Category

Category	2011 Pt Demand ('000 kg)	2012 Pt Demand ('000 kg)	2013 Pt Demand ('000 kg)
Autocatalyst	99.1	99.2	97.2
Chemical	14.6	14.0	16.8
Electrical	7.2	5.1	6.4
Glass	16.0	5.0	7.3
Investment	14.3	14.2	23.8
Jewellery	77.0	86.5	85.2
Medical and Biomedical	7.2	7.3	7.3
Petroleum	6.5	4.8	4.8
Other	10.0	13.1	13.1
Total	**251.90**	**249.20**	**261.90**

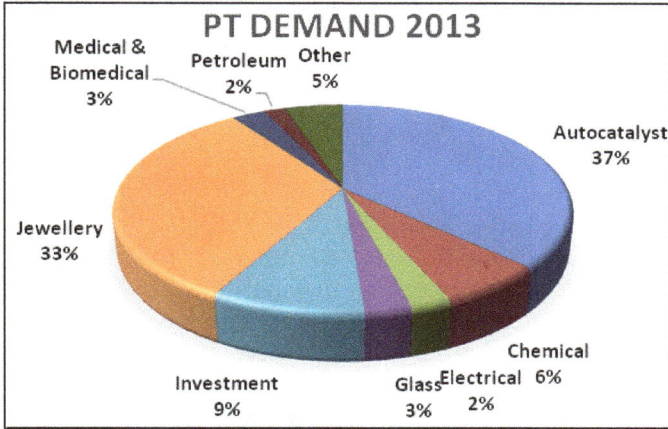

Figure 8.5.: Platinum Demand in 2013

The demand of PGMs by industry is illustrated in Figure 8.5, with about 40 per cent of platinum used in autocatalysts and about 30 per cent in jewellery.

Table 8.5: Palladium Demand by Category

Category	2011 Pt Demand ('000 kg)	2012 Pt Demand ('000 kg)	2013 Pt Demand ('000 kg)
Autocatalyst	191.4	208.5	216.8
Chemical	13.7	16.5	16.5
Dental	16.8	16.5	15.9
Electrical	42.8	37.0	32.8
Investment	−17.6	14.6	2.3
Jewellery	15.7	13.9	12.1
Other	3.4	3.1	3.1
Total	**266.20**	**310.10**	**299.50**

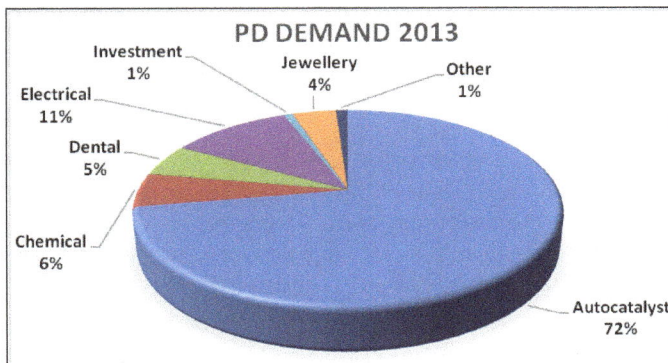

Figure 8.6: Palladium Demand 2013

There is a general increase in palladium demand. This is primarily due to advances in research to use palladium which is cheaper than platinum. Generally there is a deficit in the movement of PGM stocks. The increase in Platinum ETF is a direct result of the opening of South Africa's ABSA ETF in 2012.

The Value Chain

The production cycle of PGMs includes mining, concentration, smelting, and converting and refining as per shown below. This process ends with base metals (nickel, copper and cobalt) and PGMs (platinum, palladium, rhodium, ruthenium, iridium and gold) which are then used for making various products. Waste products from the processes include waste rock from mining, tailings (gangue which is in association with the PGMs) from concentration, slag and waste gases (mainly sulphur dioxide) from smelting and refining.

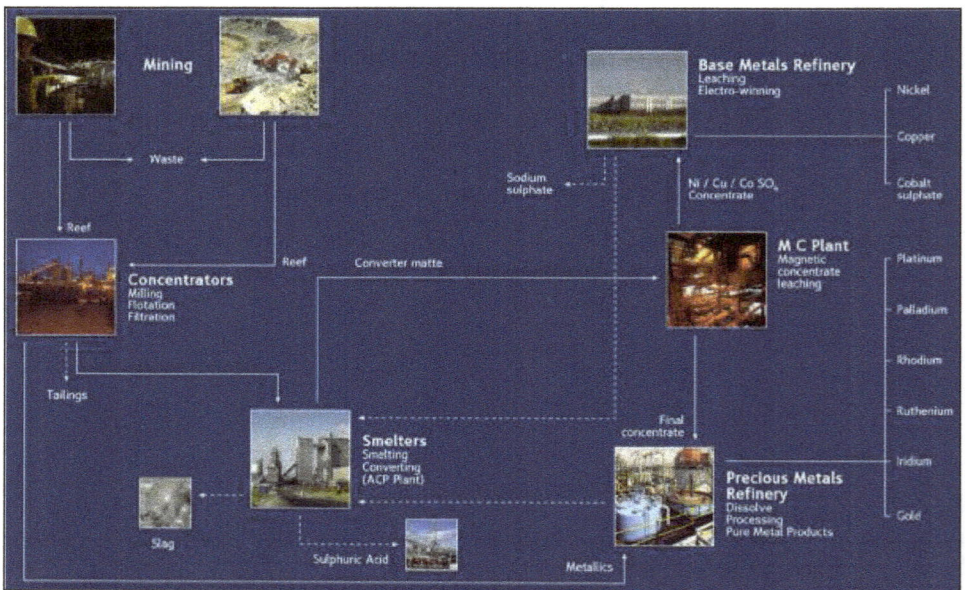

Figure 8.7: PGMs Mine to Metal

The establishment of the PGM mining operations occurs after many years of extensive pre-production investment. The road from mine to metal involves four distinct steps as outlined below:

☆ **Mining:** which is either open cast or underground. On the average the PGM grades on the Great Dyke average 3.5gpt (4E). From a tonne of ore mined and fed to the concentrator expected yield is 2.90g (4E) or 0.09 oz. after taking processing losses into account. In simple terms to produce an oz. of PGMs requires about 11 tonnes of ore.

☆ **Concentration:** treating ore involving crushing, milling, flotation and filtration to produce a PGM concentrate which also contains base metals.

On the Great Dyke concentrate PGM grades are normally greater than 65gpt (4E).

☆ **Smelting:** production of matte using high temperature in electric furnaces.

☆ **Refining:** Conversion of matte to base metals (nickel, copper and cobalt) and precious metals (platinum, palladium, rhodium, gold, iridium and ruthenium)

The Value Addition Debate

It is agreed that the country Zimbabwe needs to add value to its PGMs. The question is how far down the value chain should to go? After attending the Zimbabwe Mining Indaba 2013 where regional cooperation was mentioned another aspect also comes to mind. Which is the most ideal hub for beneficiation? In order to succeed it calls for strong regional ties and political will and commitment.

Zimbabwe should take a long term view on value addition. This should be considered as a long term project and not a populist measure. Proper planning, commitment and adequate resourcing are required. This will guarantee that value is not destroyed during the journey.

Value addition of PGMs comes with a myriad of benefits, some of which but not limited to are:

☆ Employment generation

☆ Improvement of infrastructure

☆ Technology transfer

☆ Development of downstream industry

☆ General uplifting of the standard of living for many if applied without favour and preference

☆ Huge contribution to government revenue

☆ Potential to develop research centres that could develop into centres of excellence

Value addition is only worth the effort, provided if the finished product is readily accessible to the market and consistency of supply and quality can be guaranteed. In the simplest terms, this involves refining in Zimbabwe and producing all the goods highlighted above from autocatalysts to jewellery in Zimbabwe itself. In driving the PGM value addition agenda a number of pre-requisites must be present.

☆ The infrastructure to sustain the PGM industry. Zimbabwe must satisfy itself that the required infrastructure is in place:

☐ Power and water. The industry is a huge consumer of these commodities

☐ A good road and rail infrastructure to bring in raw materials and export of goods

☐ Good communication infrastructure

❑ Reliable and competitive air transport system to move the finished high value products to the markets

☆ Local market to absorb a specified quota of the finished product

❑ Develop downstream industry such as vehicle manufacturing, biomedical equipment manufacturing, jewellery manufacture and electronics industry

❑ Raise the average household's standard of living to be able to afford the goods produced. China and India have huge populations that have a high demand for white metal jewellery!

☆ Technology and skills

❑ The technology has to be imported and collaboration with owner of technology is required

❑ Skill biased technology in different fields to be imported. A massive drive on training is required and possibility of attracting local skills in the diaspora. This is a long term, goal.

☆ Legislation framework and Governance: Must be consistent and predictable to allow for investment in the industry and ease of exporting products. Any investor requires security of tenure.

It is important to note that a workable solution may entail reserving a quota of the minerals for local use/conversion, which should increase with growth of the local industry.

Opportunities and Challenges

The development of new PGM mines is controlled by few a major mining houses, who also control the PGM stocks, and thereby, drive metal price volatility. Uncertainty in the mining industry in South Africa is stifling investment and rhetoric in Zimbabwe is shying investors away. South Africa and Zimbabwe with their PGM-dominant reserves offer the opportunity to control production, making them the preferred source of PGMs. Thus, these two countries need to be serious players. This is in no away condoning poor business practices.

The current 5 per cent or 6 per cent contribution to the world supply makes Zimbabwe an insignificant player in the PGM landscape. It is easy to substitute this source with another. However, Zimbabwe has ample reserves to improve this situation. At the moment Zimbabwe offers the cheapest oz. per tonne ore mined because of accessibility of its ores, efficiency and good work ethics.

New Zealand is embarking upon a vigorous exploration programme for its PGMs and could become a significant contributor, with the added advantage of being close to the key markets.

Zimbabwe is likely face some hurdles in its endeavour to manufacture various PGM products for a number of reasons. External auto manufacturers may have to confront higher autocatlysts prices as finished products, which normally incur higher tariffs. Zimbabwe products could find stiff resistance to enter external markets as the host countries impose protectionist policies to protect their established industries.

Smart partnerships will have to be established. Unfortunately, in this game we are not the authors of our own destiny. Globalisation must the accepted and embraced wisely.

There has been a trend to replace platinum with the cheaper palladium in autocatalysts and research continues in this regard.

Conclusions

Zimbabwe hosts sizeable reserves of PGM-dominant ore deposits. Currently, the country supplies about 6 per cent of the world market in the form of PGMs that have undergone the initial stages of beneficiation (concentrate and matte). There is a case for Zimbabwe to go a long way in the PGM value addition chain for the associated benefits it brings to the country. This is a long term project, which demands clarity of policy and commitment on part of Government. It is a national agenda requiring win-win collaborations with established players in the industry to develop skills, technology and markets. A road map to take us to the end goal is required.

For Zimbabwe to establish it self a major global player an appropriate infrastructure must be in place and should focus on developing the downstream industries to absorb some of the products. Regional initiatives will need to be taken into consideration. Zimbabwe cannot drive this agenda in isolation. PGM value addition should not be looked at as the only cure for Zimbabwe's ailing economy but diversification of all its natural resources and growth of the industry base also play an important role.

For PGM mining to be sustainable Zimbabwe needs to be actively involved in promoting use of the precious metals and finding more uses through research and development. It is undeniable that Zimbabwe has a commodity that is in demand. Direct foreign investment will play a significant role in growing the industry. The statement of Ivan Glasenberg, CEO of Glencore on 2 September 2014 on investment in South Africa sums up what investors expect, *"The country must make returns to entice us to invest. In return we invest a lot of money, we employ a lot of people, pay royalties and taxes and the government gets an on-going long term benefit."* Thus governments must create conditions conducive to investment and be efficient in collecting revenue. Glencore invested between 178-238 million US dollars in its operations in South Africa during the past three years.

The Platinum Group Metals Association (IPA) founded in 1987 has dedicated its work in promoting the use and knowledge of Platinum Group Metals. On its twenty fifth anniversary it presented 25 prominent and promising applications using PGMs:

1. Aircraft Turbines: the series of fans and compressor blades in civilian air fleets need to withstand high temperatures (greater than 1500°C) and corrosion and use platinum-aluminide coatings.
2. Autocatalysts: automobile catalytic convertors to minimise pollution from vehicles

3. Cancer Drugs and Treatment: active ingredient in chemotherapy drugs and in radio-active implants for radiation therapy

4. Ceramic Capacitors: Multi-layer ceramic capacitors (MLCC) store energy in electronic devices

5. Computer hard Disks: Performance of computer hard disks reliant upon complex structure of layered materials including platinum and ruthenium

6. Crucibles: PGMs are perfect materials for crucibles where high temperatures are necessary to produce chemicals and synthetics with utmost efficiency

7. Dental: Palladium primarily used for dental restorations

8. Electrodes and Electronics: Platinum, palladium, rhodium and iridium used to coat electrodes, the tiny components in all electronic products which help establish electrical contact between non-metallic parts and control the flow of electricity.

9. Ethylene Absorber: Controlling ethylene concentration in atmospheres where fruits and vegetables are stored reduce level of spoilage and extend shelf life.

10. Forensic Staining: Osmium tetroxide used as a stain for fingerprints and DNA. It is relatively inert and does not harm the samples.

11. Fuel Cells: Platinum used as the catalyst to convert hydrogen to energy and heat

12. Glass: Platinum and rhodium find use in LCD (Liquid Crystal Displays) and plasma manufacture

13. Hydrogen Purification: Palladium used as a thin membrane allows hydrogen to permeate through, while blocking other gases

14. Implants: Platinum and iridium used in aural implants that replace damaged cochlea in people suffering from deafness

15. Jewellery: Platinum's natural white colour, associated with its enduring quality and resistance to tarnishing has offered itself to use as decoration since the 7th century BC. The other PGMs also used as jewellery as alloys with platinum.

16. Military: Electronic equipment has platinum wiring and coating on circuits which are critical for functionality

17. Neuromodulation: devices used to treat chronic pain and diseases such as Parkinson have platinum-iridium electrodes

18. Nitric Acid and Other Chemical Catalysts: Platinum used as a catalysts in the production steps

19. Pacemakers and Defibrillators: Platinum has been used increasingly as a biomaterial, because due to its inert nature does not corrode inside the body and allergic reactions to platinum are extremely rare

20. Petroleum Refining: Platinum catalysts used since the 1950s in petroleum refineries to reform naphtha into high octane blending components for gasoline

21. Photography: Platinum and Palladium used since the 1800s for photographic prints. Still used today as an alternative process in producing archival and museum suitable prints.

22. Sensors: Platinum used in variety of sensor applications *e.g.* oxygen sensors in car exhaust systems

23. Thermocouples: Platinum and palladium-rhodium alloys for three high temperature and standard grades of thermocouples

24. Silicones: Platinum used for speciality silicones.

25. Water Treatment: Palladium promising for treating groundwater containing toxic materials

It is highly probable that more and better uses for PGMs will be developed through research. Universities and Research Centres should champion this. Both government and industry need to co-operate and support the efforts of these centres of excellence.

Glossary

Term	Definition
Pt	Platinum
Pd	Palladium
Rh	Rhodium
Au	Gold
Ir	Iridium
Ru	Ruthenium
Pdn	Production
PGM	Platinum Group Metal
tr	troy
oz	Troy ounces
kg	Kilogram
gpt	Grams per tonne
4E	Four elements, Platium, Palladium, Rhodium and Gold
g	gram

References

1. Johnson Matthey Platinum 2013 Interim Review
2. U.S. Geological Survey, Mineral Commodity Summaries, February 2014
3. L.B. Hunt and F.M. Lever Availability of Platinum Metals: Platinum Metals rev., 1969. 13
4. POLARINES working paper n.35 March 2012: Fact Sheet Platinum Group Metals

5. IPA Fact Sheet: IPA's 25th Anniversary

6. Dept. of Mineral Resources Republic of South Africa. A Beneficiation Strategy for the Minerals Industry of South Africa. June 2011

7. Fuel Cell Today: The Fuel Cell Industry Review 2013

8. Extracting Value. Stillwater Mining Company 2012 Annual report

Chapter 9

The Mineral Resources and Traditional Mineral Processing Methods in Sri Lanka and Future Prospects for Advance Mineral Processing and Beneficiation

D. Sajjaha De Silva

Deputy Director (Mines),
Geological Survey and Mines Bureau,
Sri Lanka
E-mail: gsmb@gov.lk

ABSTRACT

Development of mineral resources of a country is closely linked with its economic upliftment. Sri Lanka is reasonably endowed with a number of industrial minerals such as graphite, high purity quartz, mineral sand, limestone, dolomite, clay minerals, feldspar, apatite (rock phosphate) gemstones, dimensional stones and mica. These mineral deposits except for graphite and gemstones are mined in quarries or surficial pits by open-cast method.

Clay minerals, limestone, feldspar, calcite/dolomite and silica sand which are extensively used in local industries are not allowed for export. Minerals such as graphite, mineral sand, vein quartz and mica are being exported after some beneficiation and preliminary processing. These minerals being exported earn a significant amount of much needed foreign exchange. However, the earnings from mineral exports could be multiplied if the advanced mineral processing and beneficiations techniques could be applied before exporting them. The policy makers have taken a number of steps to maximize the mineral based income and to make the local mineral industry more sustainable by discouraging the export of raw minerals and introducing concessions for value addition.

Some of the traditional processing methods adopted in Sri Lanka include manual sorting of graphite; concentration of illmenite; rutile and zircon by the gravity separation method; manual sorting of mica and its size reduction and chipping off the iron stains. Sri Lanka also has a long history dated back to the time of ancient kings in traditional mineral processing. In the past, people have recovered Gold flakes from river sediments by traditional processing method of gravity techniques. Although some advanced processing and beneficiation methods were introduced in Sri Lankan mineral sector, particularly for vein quartz and mineral sand industry, still the traditional methods dominate. The Geological Survey and Mines Bureau together with the Ministry of Environment and Renewable Energy has taken up the challenge of introducing new policies and also the task of educating miners/entrepreneurs on some advanced mineral processing and beneficiation methods. An attempt has been made in this paper to briefly describe the economically important minerals and, to illustrate the traditional mineral processing methods and future plans for introduction of advanced mineral processing and beneficiation methods, targeting value addition and increased income.

Keywords: Sri Lanka, Mining methods, Beneficiation, Mineral processing, Value addition.

Introduction

General

The Island of Sri Lanka is located at the southernmost tip of the Indian sub-continent and its landmass covers surface area of 65,610 sq. Km with maximum length of 432 km and a maximum width of, 224 km. The country has a population of 20 million of which 74.9 per cent are Sinhalese, 15.4 per cent Tamil, 9.2 per cent Moor and 0.5 per cent Burger and Malay. The literary rate is 95.6 per cent. The Island has a tropical climate. Its central hills surrounded by plateaus rise up to 2524m above sea level. Colombo is the main business hub and the commercial center. The Island is a Democratic Socialist Republic. For greater autonomy, administration has been decentralized to provincial level.

The government's industrial strategy is to transform the domestic market oriented industry to an export oriented industry. This entails attracting foreign investment, mobilizing local capital, making local industries competitive in the world market and working towards manpower and technological developments.

After the end of civil war which plagued the country for nearly thirty years, the country recorded a strong Gross Domestic Product GDP growth of 7.3 per cent in the year 2013, owing to construction and infrastructure developments. The GDP was worth US$67.2 billion in the year 2013 with industrial sector accounting for 31.1 per cent. The per capita GDP is 3280 US$ in the year 2013.

Sri Lanka is reasonably endowed with non-metallic minerals but not with metallic and energy minerals. The economic minerals of Sri Lanka include industrial minerals such as clays, mineral sand (ilmenite, rutile, zircon, garnet and monazite), silica (quartz), limestone, dolomite, apatite (rock phosphate), graphite, feldspar, mica and gemstones.

No energy minerals have been extracted yet on land or within the Exclusive Economic Zone of Sri Lanka offshore. However, the presence of hydrocarbons within western offshore area has been reported recently. Detailed exploration by CAIRN, India is in progress.

All mineral deposits in Sri Lanka except for graphite and gemstones are mined in quarries or surficial pits by open-cast method. The only underground working mines for graphite are located at *Kahatagaha* and *Bogala*. Graphite mine at *Ragedara* has also been re-opened recently.

Except for a limited participation by the government owned companies such as Lanka Phosphate Ltd. (apatite deposit at *Eppawala*), Lanka Mineral Sands Ltd, (Mineral sand deposits at *Pulmoddai*), and *Kahatagaha* Graphite Ltd, graphite mine at *Kahatagaha*, mineral industry of the country is now in the hands of private sector.

Export and Import of Minerals and Mineral Based Products

Major mineral exports of Sri Lanka are mineral sand, quartz, graphite and mica. Export income from these minerals for the year 2011 is given in the Table 9.1. Major mineral based products that earn foreign exchange are ceramic tableware, gemstones, porcelain tableware, wall tiles, floor tiles and ornamental stones. The earnings through export of these items in 2011 exceeded rupees 4-5 billions. Although Sri Lanka has a well-established cement industry, country still has to import nearly 50 per cent of its cement requirement. The total cement consumption during the year 2011 was about 5,000,000 tones. Despite that fact that there is a large rock phosphate deposit at *Eppawala,* country still spends billions of rupees on fertilizer imports. In consideration of spending on other imports on mineral raw material and mineral based products it can be concluded that there is a large deficit in mineral export trade. In order to improve this situation, export of value added products in place of raw material and development of existing mineral deposits should be encouraged.

Table 9.1: Production and Export Data for Sri Lanka Minerals (2011)

Mineral Commodity	Total Production (tones)	Export (tones)	Local Use	Foreign Exchange Earned (Rs. Value in millions)
Mineral Sand (ilmenite, rutile and zircon)	65566	80750	Negligible	1992
Graphite	3357	3324	Negligible	450
Mica	2927	2927		131
Vein Quartz	34903	33354	1500	1114
Gemstones	9154 (thousand carats)	9154 (thousand carats)		10027
Dimension Stones	6247	6247	—	180
Sea Sand	86200	14250	Bulk locally used	21
Feldspar	53337	0	Used in local industry	Save valuable
Kaolin and Ball Clay	63000	0	Used in local industry	foreign
Silica Sand	58355	0	Used in local industry	exchange for
Calcite	14674	0	Used in local industry	raw material

Contd...

Mineral Commodity	Total Production (tones)	Export (tones)	Local Use	Foreign Exchange Earned (Rs. Value in millions)
Limestone	1231209	0	Used in local industry	export
River Sand	7046431 (m³)	0	Used in local industry	
Dolomite	195000	0	Used in local industry	
Apatite	58254		Locally used	
Salt	87256	0	Locally used	

Economic Minerals of Sri Lanka and their Current Status of Utilization

Clay Deposits and Related Industries

Clay is the main ingredient in the body mixture of ceramic ware, wall tiles, bricks and tiles. Clay deposits of Sri Lanka include kaolin (kaolinite/china clay), ball clay (a mixture of kaolinite, gibbsite, vermiculite and boehmite) and brick and tile clay (a mixture of kaolinite, gibbsite, goethite and vermiculite deposits). Kaolin (china clay) deposits mainly occur in the southwestern sector of the country. Well known kaolin deposits are present in *Meetiyagoda* and *Borellasgamuwa* areas. Much of the deposits at *Meetiyagoda* has been exploited and the mining activities in *Borellasgamuwa* area are hindered by rapid urbanization. Ball clays are mainly found in flood plains of southwestern sector. Major ball clay deposits are present at *Dediyawala* in *Kaluthara* district. Surveys to find out new deposits are being currently conducted in the southwestern part of the country. One of the well-established mineral industries in Sri Lanka is ceramic industry. There are a number of local companies which produce high quality ceramic ware for export and local use.

Feldspar

Microcline (K-feldspar) deposits occur mainly in *Rattota, Thalagoda, Kaikawala* in the *Matale* district and *Koslanda* area. Among these, the largest deposit is in *Owella* estate, *Kaikawala*. Feldspar is mainly used in the manufacture of glass, pottery, vitrified enamels and special porcelain. As important ingredients of local ceramic industry, export of both clays and feldspar is not allowed.

Silica Quartz (SiO$_2$)

Vein quartz deposits of high purity (SiO$_2$ per cent >99.65) occur in many parts of the country. Significant deposits are found in the areas of *Galaha, Opanayake, Rattota, Balangoda, Mahagama, Randeniya* and *Meegahakiwula*. It is estimated that over 1,000,000 Metric Tons (MT) of vein quarts occur in these deposits. Currently, export of vein quartz is allowed subject to prescribed degree of value addition such as in the powder form or as crushed quartz. A minor percentage (~4 per cent) is utilized locally in ceramic and allied industries.

Silica Sand (SiO$_2$)

Deposits of inland silica sand are common and well known deposits occur in the areas of *Marawila, Nattandiya and Madampe*. A quite large place of silica sand occurs in the form of sand dunes in *Ampan-Vallipuram* areas. Major use of inland high purity silica sand is in the glass industry. In addition, vast amounts of silica sand are excavated from flood plains and river beds to be used as fine aggregates. Use of dredged sea sand for construction purposes has been encouraged. Nevertheless sea sand is mainly used as a filling material

Mineral Sand

Mineral sand is a mixture of refractory heavy minerals such as ilmenite, rutile, garnet, zircon, monazite together with some silica sand. Usually, the deposits appear as stretches of black sand along the beach or occur as raised beaches. In Sri Lanka, well known beach sand deposits are present at several locations of northeastern and northwestern coastal stretches. Although the mixtures of these minerals are not uncommon, it is only at certain regions they are sufficiently concentrated for economic exploitation. Well known deposits are present at *Pulmoddai, Nayaru, Koduwakattumalai and Tavikkalu.* At *Pulmoddai* deposit, 70-75 per cent of the black sand is ilmenite, 8-10 per cent is zircon, 6-8 per cent is rutile and about 0.5 per cent is monazite. Garnet concentrated sands are reported from *Dondra* head and *Hambanthota.* It is estimated that that over 12,000,000 MT of mineral sand occurs along the beaches and in raised beaches of the island. Currently, Lanka Mineral Sands Ltd. (former Ceylon Mineral Sand Corporation) has been extracting ilmenite and rutile using magnetic and gravity separation techniques. Minerals thus separated are being exported without much value addition.

Limestone

Sedimentary limestone belonging to Miocene age extends from *Puttalam* to *Jaffna* Peninsula along the northwestern coastal belt of Sri Lanka (see Figure 1). Limestone is well exposed at several places including *Aruwakkalu, Mannar, Pooneryn* and *Kankasanthurai.* Material exposed at these localities is mainly Calcium Carburet ($CaCO_3$) and suitable for cement industry. Currently, limestone excavated at *Aruwakkalu* fulfills the $CaCO_3$ requirement of local cement industry. Export of limestone is not allowed.

Crystalline Limestone (Marble)

Chemically crystalline limestone (marble) is $CaMg(CO_3)_2$ with variable CaO/MgO ratio. Magnesium rich varieties (dolomitic marble) are used as fertilizer for long term crops and scrubbing powder while calcium rich verities are burnt to produce lime in place of coral based lime products. Thick bands of marble occur in the Highland Complex of Sri Lanka especially in *Digana* (*Kandy*), *Mathale, Dambulla, Naula, Bakamoona* areas and in the central highlands. Export of marble is not allowed.

Apatite (Rock Phosphate)

A deposit of rock phosphate with proved reserves of 24 $\times 10^6$ MT and another 15 $\times 10^6$ MT of estimated reserves are located at *Eppawala* near *Anuradhapura.* Currently, this deposit is mined by a government owned company (Lanka Phosphate Ltd.), and crushed phosphate is produced for local phosphate fertilizer requirements of long term crops. The optimum utilization of this deposit could be possible only if the phosphate rock could be converted to soluble phosphate fertilizer such as Single

Super Phosphate (SSP), Triple Super Phosphate (TSP) Di-Ammonium Phosphate (DAP) or Mono-Ammonium Phosphate (MAP). If this valuable phosphate deposit could be developed, then large sums of money drained out of the country to overseas for the import of fertilizer could be saved.

Graphite

Graphite mining in Sri Lanka dates back to the time of 1st World War. There was a great demand for Sri Lankan graphite both during the periods of 1st and 2nd World Wars mainly because of high purity. A large number of shallow pits and few major mines were in operation during these periods. Of these, only *Kahatagaha* and *Bogala* Mines are in operation at present. There is potential for establishment of graphite based industries such as manufacture of crucibles, carbon brushes, refractory bricks, electrodes, paints and lubricants. Unfortunately, bulk of the graphite mined in Sri Lanka is being exported in raw form.

Iron Ore

There are three (03) known metamorphic iron ores and few scattered supergene deposits in Sri Lanka. Metamorphic deposits are at *Panirendawa, Seruwila* and *Buttala*. These deposits are very small in size in comparison with iron ores in other parts of the world. However, in view of increasing demand for iron there is revived interest in these occurrences.

Gemstones

Sri Lanka has long been renowned for its gemstones. Perhaps nowhere in the world are so many gem varieties occur within such a small area of land. A variety of gem minerals from corundum, chrysoberyl, beryl, topaz, tourmaline, garnet, spine and quartz families are found in alluvial or colluvium deposits or in-situ. Rare stones such as Sinhalite, Ekanite and Taaffeite are unique to Sri Lanka. Traditional gem mining area was *Sabaragamuwa (Rathnapura)* but *Okkampitiya, Elahera*, Horton Plains and many other areas within the Highland Complex of the country are also now known for occurrence of precious and semi-precious stones.

Mica

The main types of mica found in the country are phlogopite, biotite and muscovite. Important commercial types are phlogopite and muscovite. Phlogopite mica is found in the areas of *Thalagoda, Madumana, Talathu-Oya, Badulla, Maskeliya, Madugoda, Udumulla, Naula, Haldummaulla* and *Kebithigollawa*.Sheet mica is mainly used in electrical appliances and in electronic industry. Scrap mica is usually ground and then used as filler-material in plastics and paint industry.

Hard Rocks Suitable for Dimension Stones and Construction Material

Recent rapid infrastructure development in the country has created a great demand for road metals, rock boulders rock aggregates. Since most parts of the island are underlain by high-grade metamorphic rocks there are numerous hard rock exposures that could be harvested for dimension stones or quarry material such as road metals or rock boulders. Since haphazard quarrying could lead to many environmental and social issues such activities have to be monitored and regulated.

However, this industry has managed to supply sufficient quantities of construction material required for the development efforts and earn valuable foreign exchange by export of valued added products.

Existing Mining and Beneficiation/Processing Methods

Mineral processing is the value addition exercise, through which the low-grade mineral deposits, industrial waste etc are processed to recover the valuable constituents and render them marketable.

An attempt is made to describe the Beneficiation/Processing Methods that are being adopted by the various industrial sectors for local utilization or export for different mineral methods. Special attempt is made to suggest the advance mineral processing methods to be adopted to earn higher revenue by value addition.

The Traditional Mining and Processing Methods for Mineral Graphite

Sri Lanka is renowned for high carbon content graphite. The best known areas for graphite are confined to the Central, Sabaragamuwa, Southern, North Western, and North Central Province of the Island.

Graphite has a long history dating back to 160 years. A high quality micro crystalline vein material was produced and the highest exports of graphite were recorded during the War years, in 1916 and 1942.

Very primitive mining methods and processing methods had been used during the peak period of graphite mining owing to the prohibitive high capital cost, lack of technology and mining professionals and also eagerness to earn quick money. Where the outcrop was located, digging started and the pitting followed vertically following the vain. Generally, the veins maintained a dip very close to the vertical.

A large number of shallow pits had been sunk in the weathered rock or top soil to produce graphite at low cost and fairly in large quantities without much regard to ground control or ground preservation. This however, caused extensive damage to the ore body and environment and consequently deep seated reserves could not be mined but left behind. Therefore, very large reserves of graphite may have been left untouched, partly due to this and difficulties faced by the early graphite miner in dewatering the mine.

Major problems faced by the miners during early days of graphite mining were de watering, haulage of material, drilling and blasting and provision of adequate ventilation.

Most of the mining and processing methods adopted were primitive in nature and the majorities were pit mines operated by manual means. Sometimes hand operated winches were used for hoisting in shallow pits. At the point of time when the miners were unable to tackle the problems related to dewatering, pit wall fixing and hoisting, they were compelled to abandon pit and sink and adjacent pit at a location along the strike. The same pattern repeated at the critical depth. This gives rise to the fact that still there is a vast potential of untapped resources beyond a certain depth awaiting future extraction.

Progress Made with Advance Methods

With the closure of large number of graphite pits as a result of loss of demand for graphite in the international market after Second World War and emergence of China and Madagascar in the graphite market scenario, a few mines namely, Bogala mines, Kahatagaha and Kolongaha Mines survived by adopting modern technology available at the time, namely:

☆ Using steam pumps and later compressed air operated pumps for dewatering,

☆ Using steam hoist for hoisting a minor reaching to greater depths and usage of compressed air hoists followed by modern electric hoists,

☆ Using compressed air operated rock drills enabling speedier tunneling and sinking practices,

☆ Ground control through the adoption of the Cut and Fill method at Bogala underground suited to the strength characteristics of the rock. At Kahatagaha and Kolongaha mines, open stopping could be successfully adopted due to greater competency of the rock.

☆ Sri Lanka vein graphite is of the highest quality natural graphite in the world. A common feature of graphite mineralization is its occurrence in the form of steep dipping veins, with widths varying from 20cm up to 2meters. Sri Lankan Graphite consist of,

☆ High purity Carbon content over 99 per cent with low ash content

☆ Nearly 95 per cent of Run of Mines (ROM) above 90 per cent Carbon

☆ Occurrence as needles, lumps, and flakes

☆ Deepest high carbon grade occurrences are at 400 – 600m depths.

The main customers for graphite are in Japan, UK, and USA. The Graphite is being marketed according to the carbon content and particle size.

Figure 9.1: Spiral Classifiers Used in Upgrading of Graphite

Bogala Graphite Lanka PLC engages in the mining, separation, refining, treating, processing and preparation and sale of graphite. The Company offers expandable graphite, graphite dispersions, and graphite parts. Its products are used in various applications, including pencils, crucibles, fireproof products, brake linings, carbon brushes, powder metallurgy, self lubricating sintered parts, energy systems, batteries, and Li-ion/Li-polymer.

Bogala graphite Lanka PLC has invested in upgrading the discarded low grades by flotation techniques and downstream products for lubricants.

Flotation is used to increase the grade of the production. "GalKatu" (G/K 30 35) is used as feed materials which contain carbon percentage of up to +90 from 30 – 35 per cent.

The size of the feed used in flotation plant is -5 mm and this was first introduced into a rod mill. After getting required size using a rod mill, spiral classifier and vertical ball mill it is fed to the rough cells. Tailings are removed at the rougher. Froth of the rougher cell is fed in to the cleaner and that in the cleaner is fed to the re-cleaner for further purification. Product of the re-cleaner is sent to the wet bagging tray, where the wet graphite is filled in to bags. Those bags are put in to the centrifuge de-watering machine which reduces the moisture content up to about 20 per cent.

Next the fed containing 20 per cent moisture is sent to the dryer. Feed rate to dryer is 0.855ton/hr and the product rate is 0.6ton/hrous, from this, the final product with 0.2 per cent moisture content is taken.

Introduction of Forth Flotation

Flotation is undoubtedly the most important and versatile mineral processing technique. Froth flotation has enabled the mining of low grade and complex ore bodies which would have otherwise been regarded as uneconomic endeavor. In earlier practice the tailings of many gravity plants were of higher grades than the upgraded ore in many used modern flotation plants.

Figure 9.2: Graphite Upgrading   **Figure 9.3: Graphite Upgrading**

Flotation is a selective process and can be used to achieve specific separations from complex ores. With using flotation cells, one can remove the other particles from ore as tailing.

Principle of Froth Flotation

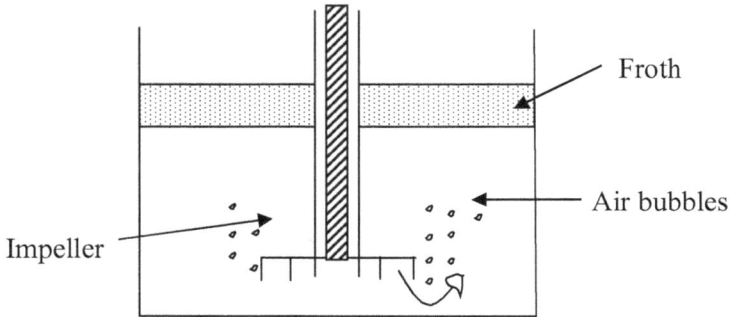

Figure 9.4: Diagram of a Flotation Cell

The theory of forth flotation is complex and is not fully understood. Forth flotation utilizes the differences in physical – chemical surface properties of particles of various minerals. After treatment with reagents, such differences in surface properties between the minerals within the flotation pulp become apparent and for the flotation to take place air bubble must be able to attach itself to a particle and lift it to the water surface. The "Ekofol" is the major chemical used as activator, collector and agent.

The process can be applied only to relatively fine particles. In case if the particles are too large then the adhesion between the particle and the bubble will be less than the particle weight and the bubble will not be able to bear the weight of the particle and therefore drop it.

The pulp enter to the bottom of the cell and is drawn up into the impeller by the suction due to its rotation and it is thrown out into the zone of agitation in the lower part of the cell. It rises up the sides of the cell and is drawn down the center into the zone of agitation again, circulating in this way during its passage through the cell. A stream of air is sucked into the impeller through pipe and is broken up into a cloud of air bubbles at lower zone. The volume of air admitted to each cell is regulated separately by means of a valve on the inlet pipe. Mineral – loaded bubble rise through the quieter upper zone and collect on the surface of the pulp as a froth which is scraped into the concentrate launder by revolving paddles. The pressure of coming feed causes pulp to flow out continuously through a slot in the back of the cell and over an adjustable weir, from which it passes down through the transfer passage to the bottom of the next cell.

It is drawn up through the impeller into the cell as before and the same procedure is repeated. The pulp passes through the line of cells in this way until substantially all of the valuable mineral has been floated off. More or less barren tailing being discharged from the last cell, the air valve and adjustable weir enable the volume of bubbles and the depth of froth to be controlled separately in each cell.

The cells have no weirs, and they are all interconnected by means of a large slot in each partition between adjoining cells. Each cell has a transfer passage with its opening in the bottom at the base of partition and the discharge under the impeller of the next cell. The suction of the impeller draws pulp through the transfer passage from one cell to the next, mixing it at the same time with the air blown in through the air inlet pipe. As the volume of pulp transferred in this way is normally greater than that of the incoming feed, the excess pulp passes back through the slot in the portion. The pulp is thus circulated through the impeller during its passage from the feed to the discharge end of the machine. A weir at the tailing discharge governs the depth of the pulp along the whole length of the machine and the depth in each individual cell has therefore to be regulated by altering the height of the over flow lip by means of removable metal strips.

Impellers are made of moulded rubber. The bottom of each cell is protected from wear by steel liners. The impeller shafts are driven by vertical motors through V-belts and grooved pulleys.

Recovery Formulae

In order to keep proper metallurgical control of a flotation plant, regular periodical samples of feed, concentrate and tailing must be taken and tested and, the feed tonnage and weight of concentrate must be accurately estimated. The tonnage of tailing is taken to be the difference between the weights of feed and concentrate. From the figures obtained, the recovery of valuable mineral can be worked out as explained below.

F = weight of feed containing f per cent valuable mineral.

C= weight of feed containing c per cent valuable mineral.

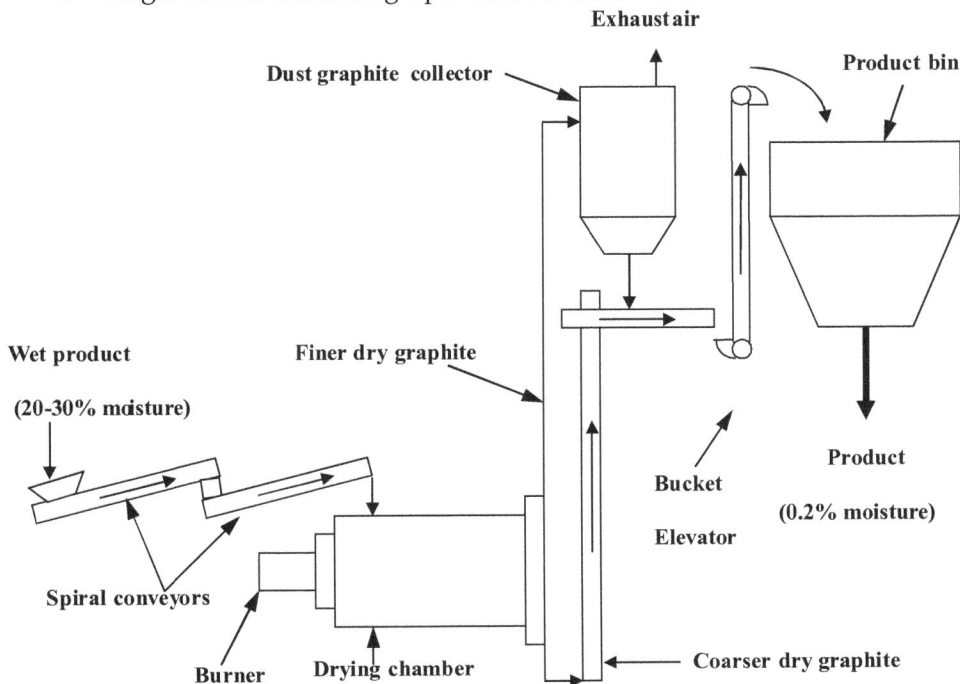

Figure 9.5: Flotation Plant Flow Sheet (Drying section)

FLOTATION PLANT FLOW SHEET

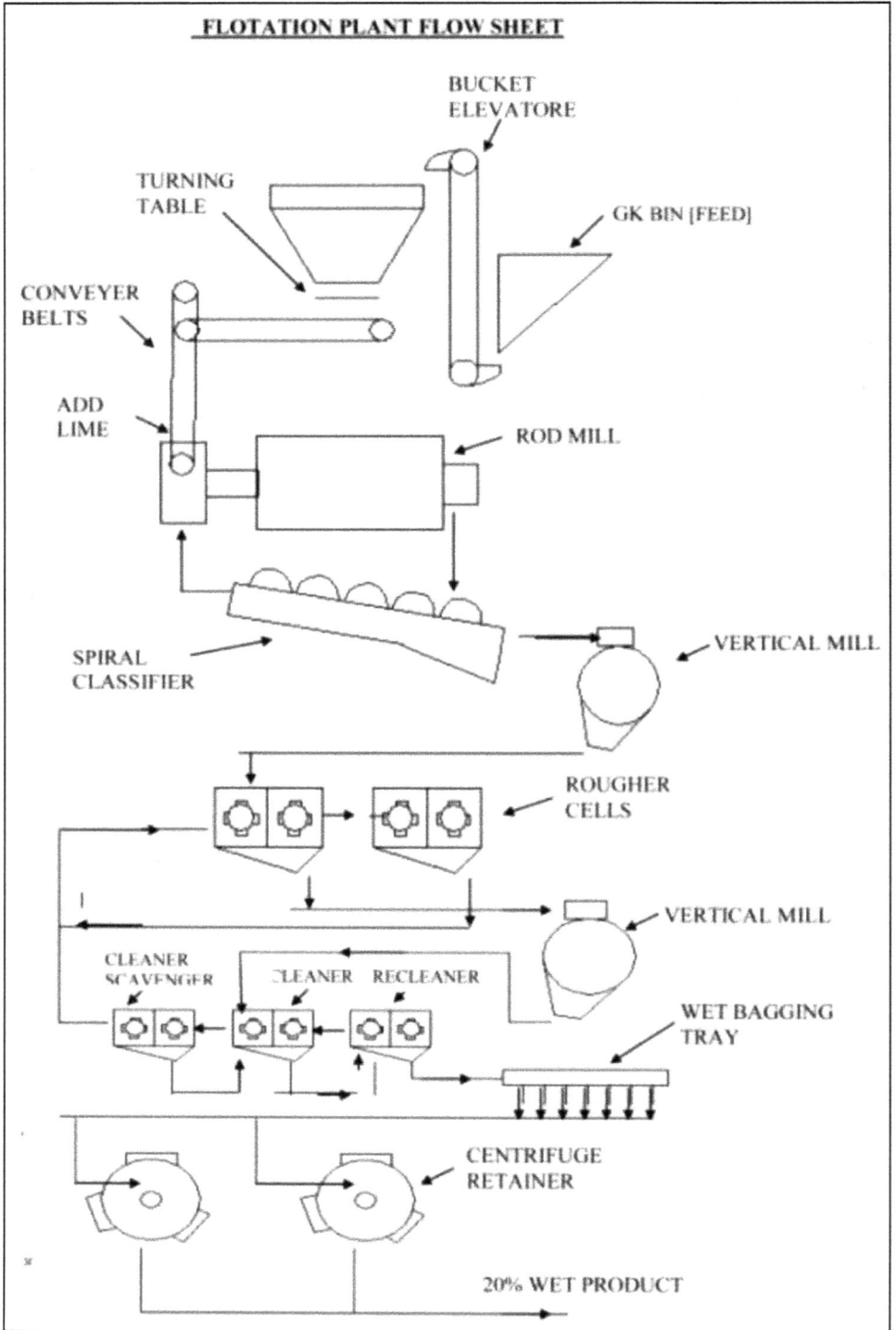

T= weight of feed containing t per cent valuable mineral.

Weight of mineral in feed = Ff/100

Weight of mineral in concentrate =Cc/100

Weight of mineral in tailing =Tt/100

Percentage recovery in the concentrate = 100*Cc/Ff

Percentage loss in the tailing = 100*Tt/Ff

Percentage recovery in the concentrate = 100C (f-t)/F(c-t)

Curing Yard

The separated sizes slabs 90 92, NO.1 OL, 85 87 and NO2 OL 80 85 from the Trommel are further upgraded by hand curing. This separated graphite is then stored in the yard in the allocated places according to its carbon content.

Curing yard is consisted of "slab yard", "curing yard" and a marking section. Hand sorting is done at this section. Graphite lumps and slabs, separated at the trommel are separated according to their grades. Curing is almost based on the surface properties of graphite *i.e.* graphite pieces are separated into different grades by observing the size, appearance and orientation of grains inside the graphite pieces. Because of this reason, sorting must be done by well skilled and experienced workers. At Bogala, there are only female workers involved in hand sorting (curing).

Figure 9.6: Female Worker Involve in Hand Sorting to different Grades of Graphite

Slab Yard

Underground production is first transported to the trammel and is separated into different sizes. Graphite pieces larger than 75mm are separated using a grizzly classifier and they are called as "slabs". These slabs are directly transported to slab yard.

Slabs containing more than 99 per cent carbon are separated and cut into pieces about 50mm in size. This is done by the skilled workers in the curing yard.

| +92 carbon | +97 carbon | +99 carbon |

Figure 9.7: Graphite Samples with different Carbon Percentages

Next, graphite slabs containing 97 per cent -99 per cent carbon are separated and cut into lumps by other workers under the direction of the supervisors of the slab yard. These lumps are called as Bogala Hard Lumps (BHL). These are harder in nature and the grains are smaller and compacted.

After separation of +99 and BHL, remainder is transported to the next step of the separation. These are called "Off Slabs".

Curing Yard

Off Slabs, NO1.OL and Off NO1.OL are sorted at the curing yard.

First, Off Slabs are broken into lumps by hammering and sieved. Next, lumps are sorted and the dust is dried using a hot plate dryer. Dried dust is taken as a feed for the mill section or flotation plant.

+97 and +95 are separated from Off Slabs; Remainder that is consisted of graphite pieces smaller than 50mm are sorted at the next step as they can't be classified as lumps.

NO1.OL is mainly separated into +97 and +92. The remainder is called "Off.1OL"

Off.1OL and remainder of Off Slabs which are smaller than 50mm are again sorted into +97, +92, Stones, Galkatu (Graphite pieces having a stone portion) and Yabora (Pieces which have been casted. These are hard like Cast Iron)

Tub Dust, NO2.OL and dust produced at the curing yard are dried using a hot plate dryer and used as a feed into mill section or flotation plant.

Traditional Processing Methods of Quartz

In Sri Lanka, silica occurs as vein quartz, silica sand, and quartzite. Vein quartz is the purest form of natural free silica (SiO_2) with SiO_2 content usually more than 98 per cent. Vein quartz deposits of extreme purity over 99.98 per cent SiO_2 are found in many parts of the Island. The best deposits occur in the Opanaike, Pelmadulla, Pussssella, Rattota, Ratnapura, Wellawaya and Galaha areas. It normally occurs on the surface as very large boulders. Some deposits may extend in depth as discontinuous veins or lenses within the crystalline rocks and they are often late formations generated from fluids migrating into dilated cracks.

Due to high purity, high resistance to thermal shock, low thermal conductivity and low thermal expansion, vein quartz is used in number of industries. Fused quartz is used in semi conductors, mould coating, metal channeling, and refractory sand in super alloys.

Due to its piezoelectric effect, quartz is suitable as resonators. Quartz also has different end use market in optics and fiber optic industry. Traditional uses of quartz are in ceramics and glassware industries. The Wellawaya quartz deposit is being now mined for establishing an industry to turn out high tech end products.

At present much of the vein quartz produced in the country is being exported to various countries without much value addition. The traditional processing methods that are adopted, including mining, sorting, braking to required lump size, chipping off of ion stains, and stock piling. The future prospects of advanced processing methods that are going to be utilized include crushing of quarts into chip form and powdering. It is expected to start quartz melting process to produce value added products of with the aid of high temperature equipments.

Vein Quartz ⟶ Heating to 1800° to 2000°C Using electricity ⟶ Fused quartz

Applications – Refractory and ceramic materials, paint, investment casting, molding compounds etc

Mining of Quartz

Quartz is mined by surface mining. A backhoe is used to remove the top soil and expose the vein. Heavy machinery is used in lager opencasts. Extraction is done by drilling and blasting while ensuring action to minimum dilution and maximum recovery. Care should be taken not to over blast resulting in excessive fragmentation. In manual work practiced in smaller mines, pick and shovel, chisels, pry bars, and dentist tools are used to extract quartz. Every measure is taken to ensure purity as the final quality is extremely important for the manufacture of processed quartz powder sold for the electronic industry.

Table 9.2: Some Production Statistics of Vein Quartz

	Production of Vein Quartz			
Year	2007	2008	2009	2010
Qty (tonnes)	35066	37196	30409	34437
Value (Rs. Million)	849.21	876.46	705.39	1020.58

Mineral Sands

The major beach mineral deposit is located in Pulmoddai, in North East of Sri lanka. Current mining is carried out by the state-owned Lanka Mineral Sand Ltd. The deposit is internationally unmatched for its concentration of over 60 per cent of heavy minerals. The deposits stretch north and south of pulmoddai for over a distance of 72 km. It is estimated that the deposit contains approximately 6 million

Figure 9.9: Quartz Powder

Figure 9.8: Production of Quartz Chips

Figure 9.10: Quartz Lamp

tones of heavy mineral sand and the resources partly replenished every year during the north east monsoon. The average composition of the sand is 70-75 per cent ilmenite, 8-10 per cent Zircon and 10 per cent futile.

Ilmenite and rutile are mainly used in the production of titanium dioxide pigment (TDP) and titanium metal. The largest market for titanium dioxide is in the paint industry. TDP is unique as a white pigment with a Refractive Index (RI) ranging between (2.55-2.7) which provides high opacity, brightness, whiteness and colour retention, tinting strength, non-toxicity and thermal stability over a wide range of temperatures. Major applications are in paints, plastics, paper, rubber, ceramics, textiles and cosmetics. Titanium metal has an excellent strength to weight ratio. Owing to its high melting point and resistant to corrosion, titanium is used mainly in aerospace industry, jet engine components and chemical plants. Zircon is mainly used as a refractory, an opacifier and in the ceramic and other industries. The bulk of world's thorium is derived from the mineral monazite containing the rare earths, cerium, lanthanum and yttrium with thoria. Thorium is also used in the manufacture of mish metal for lighter flints and gas igniters, in the ceramic, glass, electronic and electrical industries. No vertical development of the Pulmoddai resource base with a heavy mineral sand concentration has been attempted as high

Table 9.3: Export Data for Vein Quartz

Export of Vein Quartz (tones)			
Year	Country	Quantity	Value
2006	Germany	800	80000
	Japan	6004	1125828
	Korea	4166	732790
	Malaysia	620	144150
	Singapore	9782	2846720
	Taiwan	306	35044
2007	Japan	16197	4164887
	Korea	6260	882284
	Taiwan	22	6600
	Malaysia	1040	
	Singapore	7678	2283600
	Switzerland	4	609197
2008	Germany	407	52820
	Japan	21489	5116923
	South Korea	5910	903750
	Malaysia	437	114100
	Singapore	5700	1771650
	Taiwan	108	20934
2009	China	108	10800
	Germany	281	28720
	India	43	4320
	Japan	9056	2128352
	Korea	10838	1602390
	Malaysia	400	109715
	Singapore	6140	2127440
2010	Japan	13860	4336917
	Korea	7920	1183000
	Malaysia	680	207350
	Singapore	9460	3196990
	Taiwan	1	300

grade sands having over 60 per cent heavy minerals are being processed which invariably gives a high margin return on operation. Outside the pulmuddai area exploitable deposits of beach mineral sands are confined to Induruwa and Beruwala beaches rich in monazite, zircon and garnet, mouth of the Kelani Ganga (ilmenite), north of Negombo (ilmenite), Kudremalai point, south of Mannar (ilmenite and

Table 9.4: Chemical Composition of Representative Ilmenite and Rutile

Cons.	Ilmenite	Rutile
TiO2	52.53	97.53
Fe2O3	25.20	0.84
FeO	17.55	
Cons.	Monazite	Monazite
ThO2	9.51	8.65
Ce2O3	28.7	27.25
La2O3	28.56	31.08
Y2O3	1.05	0.95
P2O5	28.91	27.50
Cons.	Zircon	Zircon (crude)
ZrO2	66.40	26.53
Sio2	32.49	38.01
Fe2O3	0.187	1.28
TiO2	0.73	11.25

monazite) and along the beach in the Dewinuwara and Hambanthota areas (very high concentration of garnet (15-20 per cent). These are mainly seasonal deposits. Garnet is used as an abrasive material. The shelf area of Sri Lanka is therefore a vast unexplored region worthy of further systematic investigation. The investigation carried out in 1988 by Iluka Resources Limited of Australia has revealed that significant reserves of mineral sand are present in the North Western coastal belt around Puttalam. The estimated amount is about 1.07 billion tons of sand with 7-8 per cent heavy minerals, mainly consisting of ilmenite.

Mining and Processing of Mineral Sand

Mineral sand is mined by excavator/front loader tractor combination. In beach minerals bearing sand is manually collected, using shovels, spades, basket, and front wheel loaders. Beneficiation consists of removal of foreign trash and sea shells by

Figure 9.11: Mineral Sand Occurrence at Pesalai, Mannar District

Figure 9.12: Mineral Sand Occurrence at Lankapatuna -Trincomalee District

screening, separation of magnetic and non-magnetic fraction and magnetic fraction such as magnetite and ilmenite. Separated magnetic fraction after dewatering is warehoused for shipping. Further processing of the non-magnetic fraction involves separation into conductive and non-conductive fraction. Further purification carried out by high tension separators.

The entire beneficiated mineral sand produced from the Pulmuddai deposit is being exported in bulk, whereas only a small amount is used by the local industries.

The beneficiation of beach sand involves multi-stage processing by gravity, magnetic, electronic and flotation techniques. In recent past there have been considerable R&D efforts to develop efficient equipments in these areas for enhanced

Figure 9.13: Mineral Sand Mining and Processing at Pulmuddai Plant of LMSL

recovery and selectivity during separation of minerals. Gravity based equipment plays an important role in separation of beach sand minerals.

Since fifties only separation of mineral sands (ilmenite, rutile and zircon) has taken place at the Pulmuddai mineral sand processing plant. Processed Ilmenite, rutile and zircon sands are exported.

Value Addition to Ilmenite and Rutile

A substantial high foreign exchange could be earned by exporting ilmenite, rutile and zircon sand after increasing the concentration and also after grinding to smaller sized pedicels.

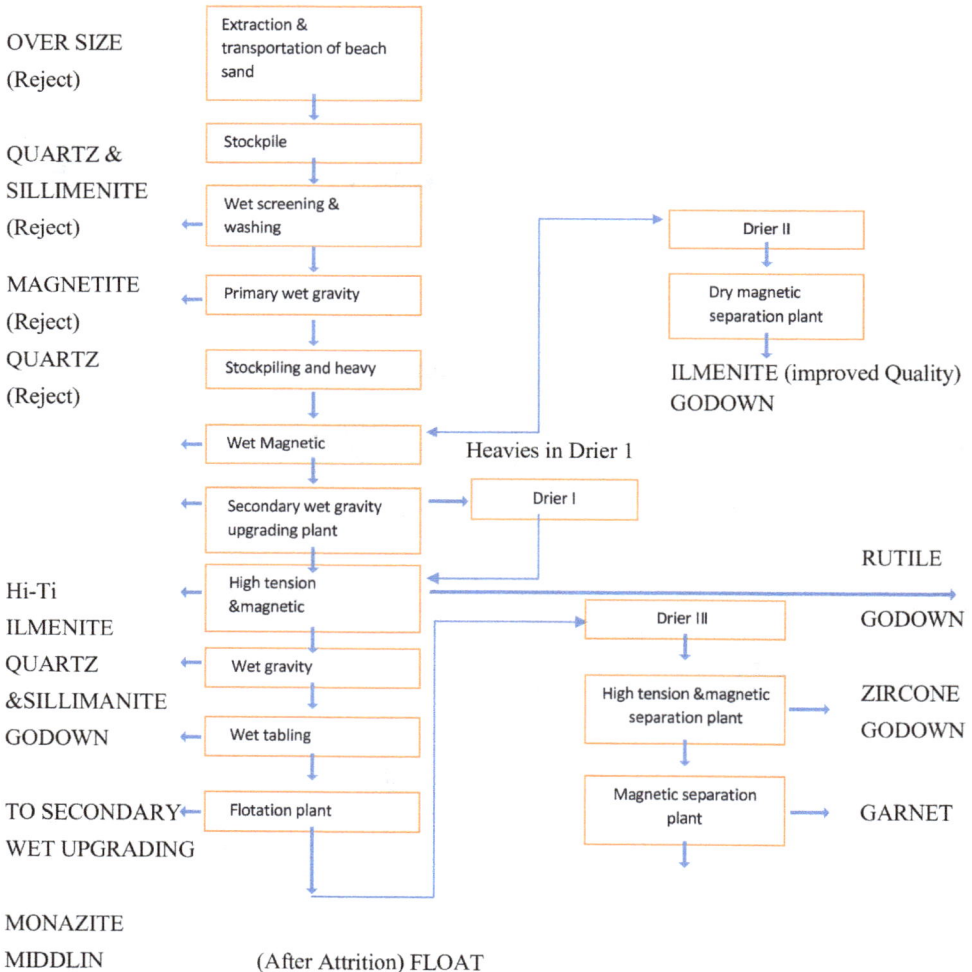

Figure 9.14: Schematic Flow Chart Proposed for the IRZ Plant–Pulmoddai

In technologically developed countries, the metal Titanium is extracted from the minerals, ilmenite and rutile. Due to its high strength, lightness, ability to withstand high temperatures and resistance to acids, it is used in the metal parts of airplanes and jets, war tanks, bone parts and propeller parts in ships. Furthermore, titanium or titania $TiO2$ is used in pigment, paint, varnish, plastic and paper industries.

The production of rutile can be given as a value addition method for ilmenite. This is carried out in many countries. There are two ways of accomplishing it.

   I. By using hydrochloric acid

   II. By using sulphuric acid

According to the hydrochloric acid method about 55-65 per cent ilmenite can be improved to artificial retile of 56 per cent TiO_2.

Here the smaller sized iron oxide is completely separated from the larger sized artificial rutile.

Stages of the Process

   I. Oxidation – Burning with air in a rotary furnace

$$4FeTiO_{3(s)} + O_{2(g)} \longrightarrow 2Fe_2O_{30}TiO_{2(s)} + 2TiO_{2(s)}$$

   II. Reduction – Heating at a temperature of more than 1200°C with Brookite, Coal and Sulphur in a rotary furnace.

$$Fe_2O_3 TiO_2 + 3CO_{(g)} \longrightarrow 2 Fe_{(s)} + 2TiO_{2(s)} + 3CO_{2(g)}$$

   III. Exposure to air – Mix with ammonium chloride and air

$$4Fe_{(s)} + 3O2_{(g)} \longrightarrow 2Fe_{(s)} + 2Fe_2O_3$$

   IV. Reduction of acidity – It is possible to completely remove the remaining iron oxide by the addition of 0.5M hydrochloric acid.

   V. Sulphuric acid method – dissolve the ilmenite sand well in concentrated sulphuric acid.then the titanium is converted to titanyl sulphate and iron sulphate.

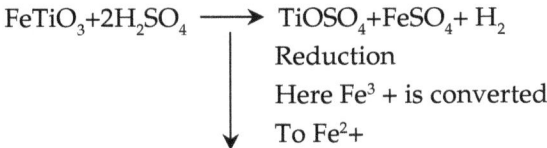

$$FeTiO_3 + 2H_2SO_4 \longrightarrow TiOSO_4 + FeSO_4 + H_2$$

        Reduction

        Here Fe^3 + is converted

        To Fe^2+

By further addition of water these can be removed as hydrates.

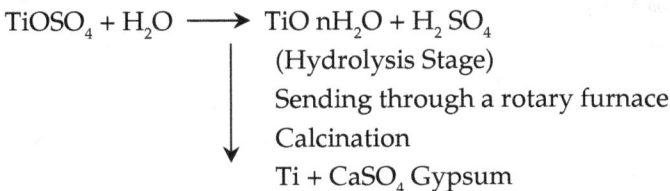

$$TiOSO_4 + H_2O \longrightarrow TiO nH_2O + H_2 SO_4$$

        (Hydrolysis Stage)

        Sending through a rotary furnace

        Calcination

        Ti + $CaSO_4$ Gypsum

Because the value addition to the minerals ilmenite and rutile is achieved by heating, concentration and filtration it involves very complex technology and initial high expenditure Sri Lanka does not have any value addition industries with mineral sands.

Figure 9.15: Zircon Separation at Dambulla Plant of AHML

Alchemy Heavy Metals (Pvt) Ltd. (AMHL) separates Zircon sand at their processing plant at Dambulla, since 2003. For this purpose non magnetic tailings are purchased from Lanka Mineral Sand Ltd., Pulmoddai plant. Processed Zr sands are exported.

Ceramic Raw Materials

Kaolin or China Clay

Economic deposits of kaolin are confined to the south west sector of the island. They are residual deposits and occur as lenses and pockets of kaolin in swampy ground. The two best known deposits are at Boralasgamuwa and meetiyagoda, where refineries have been established by Lanka Ceramic Ltd, which is the main producer in the country, At Boralasgamuwa, the percentage of kaolin in the material mined is 35-40 per cent and the deposit extend to a depth of 7 m. with an overburden of 1-2m. The Meetiyagohda kaolin deposits extend to depth of 30m. from the surface with an overburden of 2 m. thick. The kaolin content of the mined raw clay is between 35-45 per cent Super – grade kaolin is encountered where coarse grained quartz is associated with the kaolin. Super grades are presently recovered from lenses and pockets of kaolin by selective mining operation from the Meetiyagoda kaolin formations.

Refined super grades record 86-90 whiteness (fired colour at 1200° C), Fe_2O_3 and TiO_2 less than 0.25 per cent and 86-96 per cent of particles less than 8 microns in diameter. This is the grade required for export markets. A number of grades of kaolin are produced for rubber, plastics, earthenware, porcelain, refractories and other products.

Figure 9.16: Kaolin Mine at Meetiyagoda

Figure 9.17: Conveyer Belt

Figure 9.18: Filter Press

Traditional Benification Method of Kaolin

This traditional beneficiation process is normally practiced in rural areas where abundance of Kaolin deposits are available. This method is primitive and is not adopted now due to following reasons.

I. Labour involvement is high

II. Wastage of Kaolin

III. Desired particle size of Kaolin cannot be achieved

Raw Kaolin is mined by workers using mamoties and it is accumulated and collected in heaps. Kaolin is next fed to an underground cemented tank. This tank is not much deep and Kaolin and water are fed into the tank. Some workers get into the tank and get involved in trodden.

The main purpose of this trodden is to mix Kaolin and water in order to prepare the slurry. Water is continuously poured into the tank, and overflow is gradually goes into the adjoining tank as shown in the diagram.

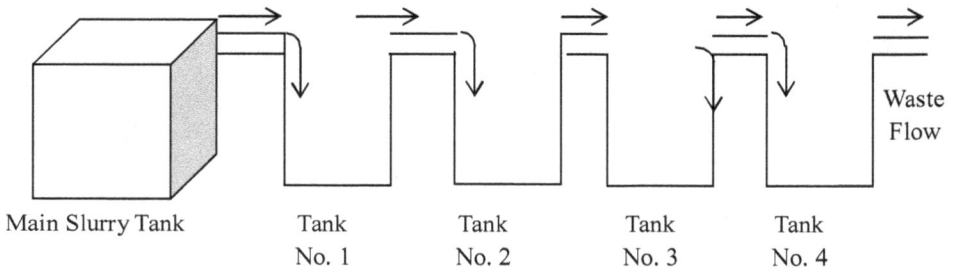

Figure 9.19: Series of Tanks are Installed for the Over Flow of Slurry

Kaolin slurry flows in this manner from tank no: 1 to tank no: 4. If the waste flow is still rich with Kaolin, additional tanks are installed in the series.

Tank No. 01

Filled mostly with coarse and fine sand with clay.

Tank No. 02

Sand content is low than tank no. 01.

Tank No. 03

Sand content is low than tank no. 02 continuously this slurry flow until waste water is free of Kaolin and sand. If it needs more purification depending on the buyer, content in tank no. 02 can be recycled by putting the slurry into main tank and adding water to overflow from main slurry tank and proceed to other tanks accordingly.

Modern Method of Benification of Kaolin fields

Mining is done by excavator, during mining if we find that the Kaolin is mixed with other impurities, it is manually sorted out. Mined Kaolin is then loaded into tippers and transported to the plant and stored near the feed hopper.

Refining Process

Using the front end loader it is fed into feed hopper, and then Kaolin is fed into a belt conveyer, Kaolin lumps goes through the crusher which crumbles into small pieces and goes to a horizontal blunger, where Kaolin gets mixed up with water, coarse kaolin mix water passes to mica drag. At this stage further separation of mica, coarse sand and kaolin take place. The balance slurry passes through sieve (40 mesh) which involves further separation of mica and fine sand.

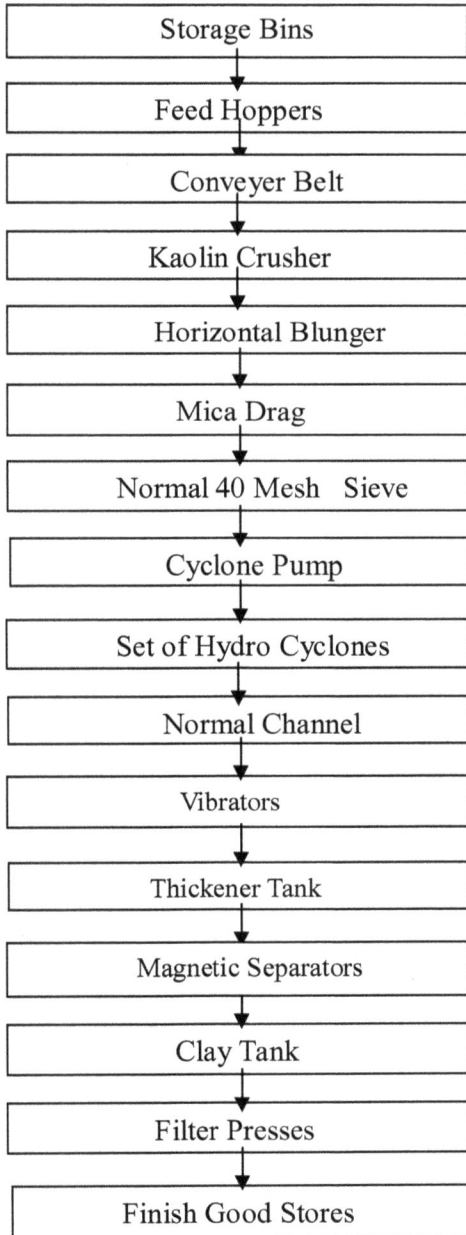

Storage Bins

Feed Hoppers

Conveyer Belt

Kaolin Crusher

Horizontal Blunger

Mica Drag

Normal 40 Mesh   Sieve

Cyclone Pump

Set of Hydro Cyclones

Normal Channel

Vibrators

Thickener Tank

Magnetic Separators

Clay Tank

Filter Presses

Finish Good Stores

Figure 9.20: Flow Chart Kaolin Manufacturing Processing

Remaining slurry is pumped into a set of hydro cyclones. The function of hydro cyclone is to separate fine sand and coarse clay particles. Overflow portion of hydro cyclone passes through a long length channel. The channel is position to a low gradient in order to decrease flow speed of slurry. The slurry is next passes through vibratory sieves. (200 mesh/250 mesh) It is then passes through a channel to thickener tank (settling tank).

The settled clay slurry passes through magnetic separator to a tank and then filter presses to reduce water percentage. The pressed clay known as pressed cake is ready for marketing. If it is needed in a powder form then the cake has to be subjected to drying and grinding processes.

Feldspar

Microline (K- feldspar) deposits occur mainly in Rattota, Kaikawala in the Matale district and Koslanda area. Among these, the largest deposit is in Owella estate, Kaikawala. Feldspar is mainly used in the manufacture of glass, pottery, vitrified enamels and special porcelain. As important ingredients of local ceramic industry export of both clays and feldspar is not allowed.

Feldspars Used in Ceramics Industry

Floor Tiles – 10-55 per cent

Porcelain ware – 15-30 per cent

Sanitary ware – 25-35 per cent

Figure 9.21: Feldspar Mine at Thalagoda

Apart from Lanka Ceramics Ltd, there are several other minor producers who supply feldspars from different parts of the country for local users.

Lanka Ceramics Ltd. Produce 2000 metric tons of Refined kaolin

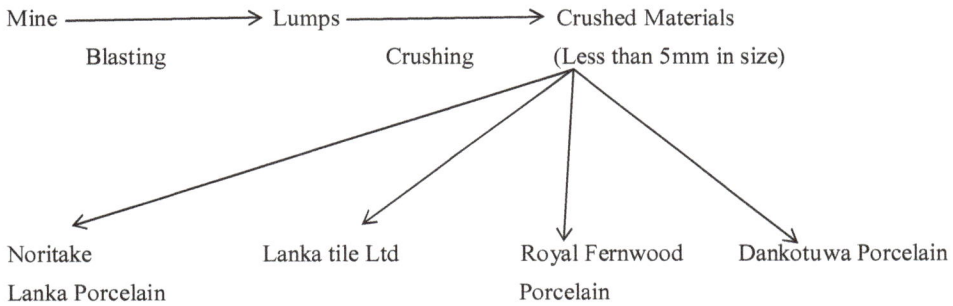

Mine ――――――――→ Lumps ――――――――→ Crushed Materials

Blasting   Crushing   (Less than 5mm in size)

Noritake   Lanka tile Ltd   Royal Fernwood   Dankotuwa Porcelain

Lanka Porcelain   Porcelain

Figure 9.22: Distribution from Feldspar from Mine to End-users

Miocene Limestone

Miocene Limestone is a very common sedimentary rock of biochemical origin. It is composed mostly of calcite. The Miocene limestone occurs in the north western coastal belt and the Jaffna peninsula. Over a large area of this belt the limestone is poorly exposed being covered by superficial deposits. The best known outcrops occur in the center of the Jaffna Peninsula, where it attains a thickness of several meters. Outside the peninsula, the Karativu, near Puttalam, where the limestone, including both pure and siliceous types, stretches from a series of low hills running northwards from Aruakkalu hills to Kudremalai point.

Miocene limestone is mainly used for cement manufacture. At present, Holcim (Lanka) Ltd. owns and operates the only existing quarry, located at Aruwakkalu, north of Puttalam with the rapid exhaustion of inland coral and shell deposits mainly confined to the south west and southern coastal of the country, Miocene limestone is the only alternative raw material available for lime production.

Figure 9.23: Mining/Blasting-Raw Material

Figure 9.24: Production of Klinker using Limstone

Feldspar Mine

Drilling & Blasting

Feldspar lumps breaking by excavator in to 6¿x 9¿ size

| Good Feldspar | + | Manual selection by chipping of feldspar |

By Tipper

| Feldspar   storage |

By Front End Loader / Tippers

| Feed Hopper |

| Jaw Crusher |

Over size by conveyer

by conveyer

| Vibrator Screen |

Over size by conveyer

| Cone Crusher |

| Storage Bins |
| Feldspar chips (< 5 mm) |

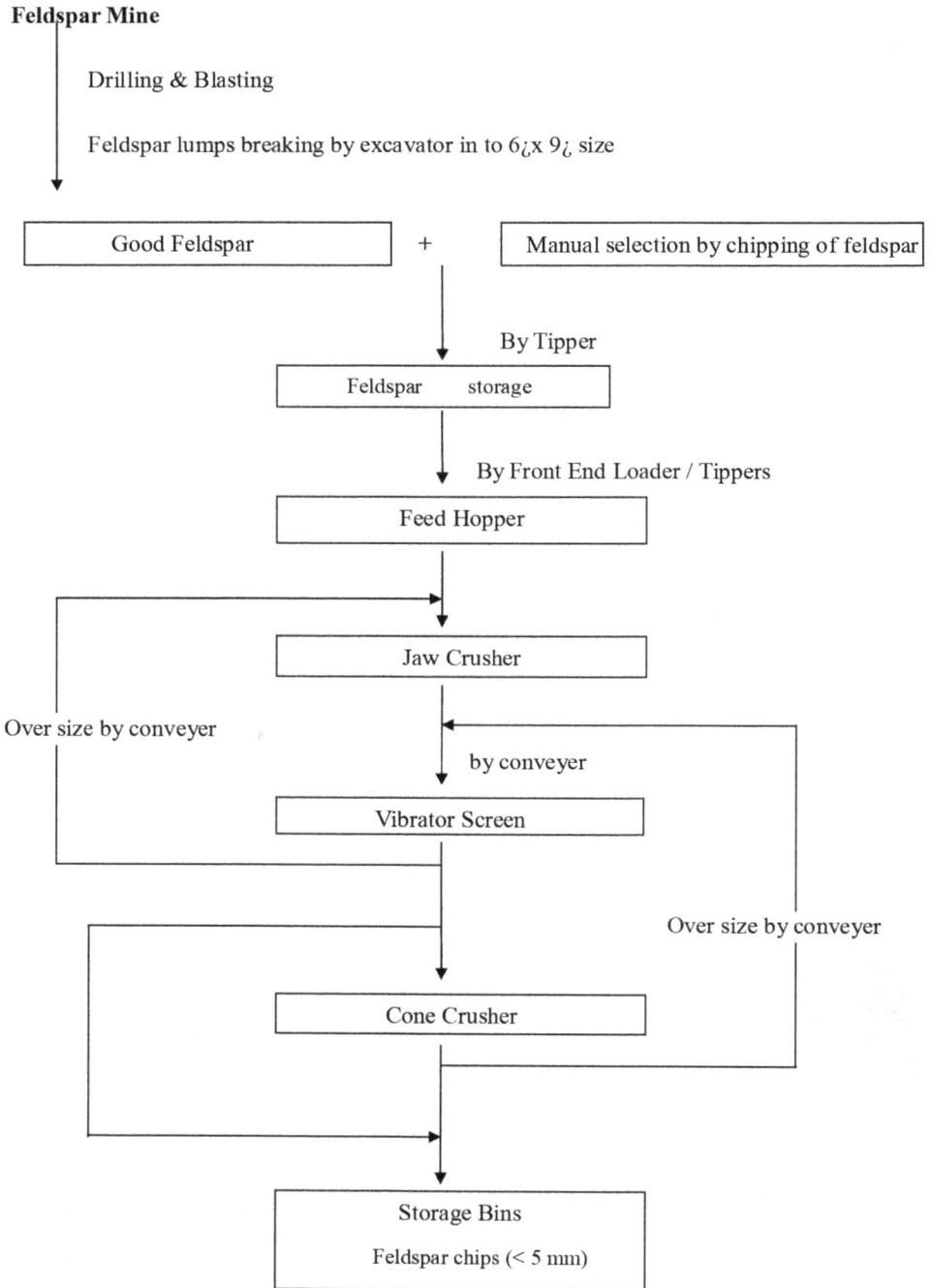

Figure 9.25: Flow Chart Feldspar Processing

Miocene Limestones are mainly used for cement manufacture.Presently, Holcim (Lanka) Ltd owns and operates the only existing quarry at Aruwakkalu, north of Puttalam.

Mica deposits in Sri Lanka mainly confined to the central hills of the country are found in the Talagoda, Madumana, Pallekalle, Talato-oya, Badulla, Maskeliya, Madugoda, Udumulla, Naula, Haldummulla, Mailapitiya, Dutu Wewa (Kebithigollawa) and Madampe areas. The important commercial types of mica aremuscovite (K_2O-$3Al_2O_3$-$6SiO_2$-$2H_2O$) and phologopite (K_2O-$6MgO$.Al_2O_3-$6 SiO_2$-$2H_2O$). Mica is classified as sheet, scrap and flakes.

Sheet mica is mainly used in manufacture of electrical appliances, and electronic industries. Scrap and flake mica are processed into ground mica. Dry ground mica is used as filler in plastics, floor coverings and paints. It is also used for lagging steam pipes and boilers. Wet ground mica is used for decorative work, in wall-paper, printing and in the rubber and other industries.

Mica

Removing Weathered Portions

Sorting Scrap Mica

Sorted Scrap Mica

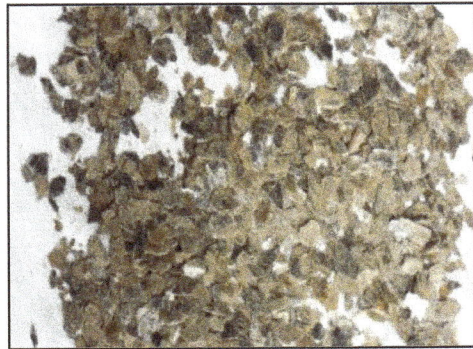

Sorted Waste Mica

Figure 9.26: Processing of Mica

Table 9.5: Mica Exports (tonnes)

Country	2007		2008		2009		2010	
	Qty. (Tonnes)	Value ($)	Qty. (Tonnes)	Value ($)	Qty. (Tonnes)	Value ($)	Qty. (Tonnes)	Value ($)
Belgium	–	–	24	12000	24	12000	–	–
China	316	95385	858	307860	991	378082	1456	556354
France	–	–	–	–	44	22000	–	–
German	–	–	–	–	–	–	100	46887
India	–	–	–	–	10	5250	–	–
Japan	1315	448700	1393	545570	1120	487295	482	213300
South Korea	–	–	–	–	–	–	–	–
Romania	0.3	150	49	29000	–	–	–	–

Conclusion and Recommendations

An attempt has been made to describe the mineral processes prevailing in Sri Lanka and highlight the present situation in context of developing and upgrading the mineral processes. Mineral are non renewable resources and utilization of such resources should be carried out on a carefully planned long term programme. Immediate steps should be taken to assess ongoing mineral processes and see what remedial measure should be taken for modernization and upgrading of advanced mineral technologies for the benefit of the country. Geological survey of Mines Bureau on long term plan should undertake a mythology to sustain and expand the existing processing method and to start new industries based on advanced mineral technology. In order to realize this objective the government should provide all possible assistance to improve the research and development facilities with special emphasis on pilot projects with the direction of Geological Survey and Mines Bureau to introduce advanced mineral technology to the country.

Acknowledgement

The author would like to convey his gratitude to organization Committee of the Ministries responsible for Mines and Mining Development, Higher and Tertiary Education, Science and Technology Department of the Republic of Zimbabwe, Center for Science and Technology of the Non-Aligned and Other Developing Countries, New Delhi, India for given me opportunity to present this paper at the 3rd International Workshop on Mineral Processing and Beneficiation at Harare, Zimbabwe.

The author extends his gratitude to officials of Bogala Graphite Lanka Ltd, Lanka Mineral Sand Ltd, Lanka Ceramic Ltd, for assistance and providing data for this paper.

The author owe his depose gratitude to Mr. Senarath Jayasundara, Chairmen GSMB, Mr. Anil Peiris, Director General GSMB, Dr. Bernad Prem, Deputy Director of GSMB, and staff of mines Division of GSMB for their valuable support and assistance given during processing preparation of this paper.

References

1. Cummins, B, SME Mining Engineering hand book, volume 2, PP 27-1-27-94

2. Dissanayake, C.B.(1982) The Geology and Geochemistry of the Uda-walawe Serpentinite, Sri Lanka,J, Natn.Sci.Coun.SriLanka 10 (1) 13-34

3. Herath, J.W. (1980) Mineral Resources of Sri Lanka Economic Bulletin No 2 (Second Edition Revised) Geological Survey Department.

4. Loveson, V.J.Misra, D.D. (2004) Sustainable Development of Coastal Place Minerals, PP 146-15.

Chapter 10

Developing Minerals Market Places and Commodity Exchange Mechanisms

Richard Tushemereirwe

Senior Presidential Advisor on Science,
Statehouse, Uganda
E-mail: richadt2002@yahoo.com

ABSTRACT

Minerals market places have historically provided the physical organization facilitating meeting of many players in the trade and thus enabling value addition. In addition, they led to evolution of Minerals Commodity Exchanges, a great pillar in wealth creation. Whereas, livestock markets were organized to give local support for the colonial enterprise, trade in minerals were never organized because they aligned for export to Europe to feed industry. Developing countries need to organize mineral trading by building minerals market places to support commodity exchange mechanisms, which in turn act as enablers of value addition.

Keywords: Minerals market places, Commodity exchange.

Introduction

What is a Commodity Exchange?

A platform where commodities and derivative products are traded. Most commodity markets across the world trade in agricultural products and other raw materials (wheat, barley, sugar, maize, cotton, cocoa, coffee, milk products, oil, metals, *etc.*) and contracts based on them. It can also be a physical market place where stocks can be graded, certified, and then traded, exchanged and converted into various financial instruments. This can be a fundamental infrastructure to realize mineral wealth.

What are Commodities?

Standardized products: graded, certified, with a traceable chain of custody, stored securely for appreciable longer period in a safe place and ultimately deliverable when required.

Evolution of Modern Commodity Exchanges

Commodity exchanges were evolved mainly to create price stability in supply of goods for industry. The Dojima rice market in Osaka Japan evolved the first modern commodity exchange in 1730.This rice commodity exchange was so popular and efficient, the government of Japan accepted taxes in form of rice. These developments by the rice merchants of Osaka were linked to parallel developments led by merchant associations in other trades. For example, a corn producer could sign a forward contract for sale of specified amount of corn at a future date and at a specified price to a processor or retailer. A speculator, with unique information and therefore with a high risk appetite, could buy this futures contract with the hope of profiting from anticipated price hikes.

Around 1840, the telegraph and the steamship were two major developments that shaped the commodities market. The steamship reduced the time of shipping significantly that non-local demand was taken into account in commodity trading. The impact of the telegraph was even more significant because information traveled faster than the commodities. The Chicago Board of Trade (CBOT), the first commodity exchange in USA opened in 1848 dealing in agricultural products, especially grain and is still the largest in the world. The US department of Agriculture sets standards for agricultural products to be traded on the exchange.

In 1864, CBOT listed the first ever exchange traded forward contracts which were called futures contract thus linking the commodities markets with financial markets. The futures market came into being to include financial instruments like currencies, bonds, and shares. Speculators provided the market with liquidity and with appreciation in prices, they profited. This also helped to create a transparent market with minimal fluctuations.

The third commodity exchange opened in London in 1877 trading in mineral ores and metals, the London Metal Exchange (LME). A year before, the opening of Suez canal had reduced the delivery time of tin from then Malaya to match the three months delivery time for copper from Chile. With continents connected by telegraph, merchants were able to anticipate time of arrival of metal cargo and were able to sell it for future delivery, on a fixed date, thus protecting themselves against a fall in prices during the voyage. LME was also born out of the need to organize the mineral stocks, once they arrived on the various shores of England, in a manner that buyers and sellers could meet in one place of trade while the stocks were securely kept to be delivered when and where they were needed.

Five commodity exchanges had emerged by the end of 1800s all connected by the transatlantic cable network: New York, New Orleans, Liverpool, Havre and

Alexandria. These commodity exchanges became global markets because what was happening in Alexandria influenced what was happening in New York – the birth of globalization.

The post-world war II period 1940 – 1970 saw many government interventions that brought down commodity exchanges. Protectionism in Europe and America, central planning in socialist countries and prohibition policies in countries like China, India and Egypt saw the collapse of commodity exchanges globally.

The 1970s saw a rebirth in commodity exchanges after collapse of Breton woods system. The US Dollar was untied from gold and the two emerged as two autonomous markets. The oil market also emerged after the crisis of 1971 because oil was now priced on market based arrangement instead of contractual agreements. Oil is the most traded commodity in the world followed by coffee.

In Africa, the Ethiopian Commodity Exchange (ECX) was founded as an innovation to solve the paradox of hunger in one region and a surplus of food in another region of the same country. In the words of ECX founder, Eleni Gabre Madhin, *"One of the things I kept seeing over and over, which I'd seen in other parts of Africa, was just how difficult it was for buyers to find sellers and sellers to find buyers, and how difficult it was to enforce the contract"*.

Learned from the Evolution of Commodity Exchanges

Most of the African countries and other developing nations are endowed with vast mineral resources, but adding value to the extent of creating end-user products from these minerals has remained unfulfilled dream. Just like in the example of Ethiopia with grains, there is abundance in one place and scarcity in another all within the same economy and political boundaries.

Whereas, many African governments have written down policies that express the desire of fully extracting value out of their mineral resources, the dynamics of international trade in minerals is structured in such a way that quick money can only be made if these countries export semi-processed minerals to the developed world. How? The developed world has more organized markets for minerals and metals while the developing countries do lack in that. Like in all other sectors, it is always an ecosystem with many players supporting each other. In biological ecosystems, an organized ecosystem is one which is balanced and self-sufficient leading to attraction of resources from neighbouring non-balanced ecosystems.

In these developing countries, there is no infrastructure to support investment in smelting, refining and trade in minerals. This is further complicated by disorganized marketing of minerals that is of no help to both governments and investors. A comparison with the livestock markets can give a clue how history and organization or lack of it can play a role in whether there is development of a sector or the whole economy. All across the developing world, colonial governments built designated market places for livestock and a systematic set of procedures were evolved for marketing and trading in livestock and derived products. These market places provide a meeting point, both physically and virtually, for players of various kinds in livestock industry – the cattle keepers, slaughterers, hides and skin processors, horns and hooves processors, and many other dealers. This is how the livestock industry was able to make a significant contribution plus generating good revenue to both the colonial and post-independence economies.

However, it should be understood why livestock markets could be so much organized and not the mineral markets. Minerals were required in Europe to feed industries but not livestock which was reserved to support the colonial enterprise locally.

Therefore, the lessons that can be learnt from the evolution of commodity markets world-wide and how we can benefit from these lessons to maximally exploit our mineral resources are many. One such lesson is that there should be organization of the mineral sector starting with physical market places for trade in minerals and related products. These physical designated market places provide the nucleus organization that radiates in the entire minerals sector to build a strong pillar for the economy.

The other lesson is that there abundance of elastic wealth far and above the mere counting of physical quantities of mineral deposits or mined stocks. This is derived wealth and can only be realized after organization in the marketing and trading of minerals.

Commodity Exchanges as Enablers of Value Addition from Minerals

Any commodity exchange is a perfect ecosystem or micro-environment providing a functional ecosystem with linkages to many other forms of wealth worldwide. Like all functional ecosystems, it has the ability to grow in an elastic manner. To start with, commodity exchanges are membership based trading associations.

Both buyer and seller at a commodity exchange do not exchange physical commodities, instead, they exchange certificates of ownership. The exchange acts as a third party guarantor of the transaction and that is important. Having a guarantor for the transaction means you do not have to beg people to pay you or chase after them. You do not have to worry about the quality, quantity, delivery or payment. The exchange guarantees all these things. This is a very big value addition enabler.

While speculators play a major role in commodities market, producers and consumers have a fair share of participation. For example, farmers sell futures contracts equivalent to their crops at a specific time if they are worried of impending price drops. Miners sell confirmed deposits in their mining area if they do not have enough capital to develop the mining enterprise.

Market places for Minerals and their commodity exchanges serve as enablers of Value Addition by offering infrastructure that serves as interface between miners, processors and dealers of mineral ores and derived products. Designated warehouses/stores to offer grading and certification services for minerals of all grades (ores, intermediate and refined grades). They offer a traceable Chain of Custody and trade enabling government to maximize revenue collection plus minimizing of risk to the buyers and sellers, ease of law enforcement and security to owners of mineral stocks, better management of environmental and human health regulations.

Minerals Commodity Exchange enables value addition in secondary ways. Convertibility with other forms of wealth is important. Mineral stocks can be converted into financial instruments including currencies, stocks, mortgages, *etc.* All tradable commodities can be cross traded on stock exchanges, financial and currency markets.

With commodity exchanges, informal speculation can be replaced with formal speculation. The minerals sector in most developing countries is trained with all sorts of criminality. Advantages of formal speculation are that it can be regulated and taxed by government.

A government policy remains unfulfilled wish unless it is enforceable. Enforcement is mainly reliant on information made available and ability to monitor continuous information flow. Market places for Minerals and Commodity exchanges enable governments to manage policy frameworks based on real data and make quick and timely interventions.

Markets for Minerals and Commodity Exchanges are also great investment enablers. They ensure steady availability of raw materials and intermediate products for secondary and tertiary processing. They offer a platform for mobilization of capital for infrastructure and crating public goods.

Most importantly, these market places act as meeting points and reservoirs for miners, processors, middlemen, and manufactures of various end-user products. The long term stability of prices and constancy in supply is achieved through these mechanisms.

Moving Forward

By government regulation, all minerals should be stored and traded in designated warehouses, stores or appropriate market places. The warehouses, stores or appropriate market places should offer the following services: traceability – documented chain of custody, grading and certification, plus accessory services like auction houses.

Government should also create enabling environment for establishment and good operation of Minerals Commodity exchanges by the private sector.

Chapter 11

Current Status of the Mining Sector and Proposed Actions for its Beneficiation in Uganda

Henry Mugisha Bazira

Executive Director,
Water Governance Institute (WGI),
Arknet Building, Plot 398 Kalerwe-Gayaza Road,
P.O. Box 23704, Kampala, Uganda
E-mail: watergovinst@gmail.com, bazirah@yahoo.co.uk, info@watergovinst.org

ABSTRACT

Uganda has a vast array and quantities of mineral and petroleum deposits. The mineral deposits are fairly distributed across the country, but the highest concentration of minerals lies in the south-western part of the country. The minerals are still underdeveloped and underexploited. This presents a significant opportunity for investment. The traditional minerals that have been mined at industry/commercial level include copper, tin, wolfram, cobalt, and limestone which experienced periods of dwindled production during periods of political and armed conflicts, but are now being revived. Gold and iron ore have been mined at artisan level. Other minerals that exhibit commercial potential include phosphates, kaolin, chromium, columbite-tantalite, beryl, *etc.*

The country discovered large quantities (3.5billion barrels) of oil, whose discovery has triggered excitement and trepidation on the opportunities and challenges that the resource presents. Government is fast-tracking the revision of existing policies and regulations to meet the demands, requirements and standards of the emerging petroleum industry. It is hoped that the revisions will also meet the demands, requirement and standards of the mining sector, because this is another growing sector. The petroleum resource alone has the potential to increase government treasury reserves by an amount equivalent to US$3.3billion annually once commercial production begins. This can be a very significant contribution to annual national budget.

There is a unique overlap of mineral deposits and conservation areas in the country that may present legal, political and development challenges and potential conflicts between conservationists and development planners, if nothing is done to address the legal lacunae and interests.

Mineral exploitation should be pursued to meet the country's broader social and economic goals, but not as an end in itself. This requires having a vision of how the resource fits in the country's economic future. For some countries, the best use could be to leave the resource untapped for a period of time, while for others, it may need to exploit the resource immediately to generate revenue and create wealth to sustain investments for growth and meet basic need of the citizens. Whatever the decision, there are practical guidelines that need to be undertaken to maximize the opportunities and minimize the challenges presented by any given mineral resource, including the following:

☆ Using the mineral resource for social and economic benefits;

☆ Government being accountable to inform public;

☆ Formulating fiscal and contractual terms that are robust to change the uncertain circumstances;

☆ Competition in the award of contracts and development rights;

☆ Careful decision-making whether or not to exploit taking into account local economic, environmental, social and political effects. Private sector led versus People centred development

☆ Having clear, transparent and accountable revenue sharing, transfer and investment based on stipulated rules, procedures and sound judgment;

☆ Using natural resource revenues for additional wealth creation social and economic growth and development;

☆ Caution in entering international trade and investment agreements/treaties.

It is important for a country to assess self based on how these guidelines are being fulfilled or complied with as she moves into the exploitation of her mineral resources.

This paper highlights the current status of the minerals and mining sector in Uganda and aims at sharing ideas that could assist the country and other mineral resource-rich countries in Africa initiate interventions that generate economic growth, enhance citizen welfare and ensure socio-economic and environmental sustainability.

The paper was prepared on the basis of the author's knowledge and experience and a review of literature about minerals and the mining sector in Uganda. Inferences were drawn from literature and the situation at hand in the country and recommendations made.

Introduction

Uganda is a natural resources-rich country, including forests, vast freshwater bodies, arable land, domestic animals, wildlife and minerals (metallic and non-metallic). While the forests, freshwater bodies, arable land, domestic animals and wildlife seem to be fairly developed and exploited, the mineral resources are generally under-developed and at various stages of exploitation.

Agriculture and other non-mineral natural resources contribute up to 80 per cent of the Gross Domestic Product (GDP), while mineral resources contribute about 7 per cent of GDP not withstanding their vast occurrence. Revenues from the emerging petroleum industry have the potential to double government income within six to ten years of commercial production and constitute an estimated 10-25 per cent of Gross Domestic Product (GDP) at its peak production.

The mining sector presents special opportunities and challenges for mineral-rich countries. The opportunities include using the mineral resources to create greater prosperity for current and future generations, if used properly. While the challenges include causing economic instability, social conflict and lasting environmental damage, if mineral and their associated revenues are used badly.

The Constitution of the Republic of Uganda makes a clear distinction between mineral and petroleum. Therefore, for the purposes of this paper, the distinction shall be maintained. However, this paper has occasionally used the term "mineral" when also referring to petroleum.

The purpose of this paper is to highlight the current status of the minerals and mining sector in Uganda and to share ideas that could assist the country and other mineral resource-rich countries in Africa initiate interventions that generate economic growth, enhance citizen welfare and ensure socio-economic and environmental sustainability.

The paper has been prepared on the basis of the author's knowledge and experience and a review of literature about minerals and the mining sector in Uganda. Inferences were drawn from literature and the situation at hand in the country and recommendations made.

Minerals and Mineral Sector in Uganda

Mineral Occurrence

The metallic mineral resource potential of Uganda is depicted in mineral occurrence (Figure 11.1). It is clear that Uganda has a vast array of metallic minerals that are almost evenly spread across the country. it is clear that there are vast deposits of Phosphates, Gold, Copper, limestone, Cobalt, Columbite-Tantalite, Tin, Wolfram, iron, Chromium and Beryl in Uganda.

The mineral occurrence was established through geological data acquisition, processing and interpretation of 80 per cent of the country. The remaining 20 per cent is yet to be completed.

Geochemical and geophysical mineral surveys of, for example, the Karuma area revealed existence of nickel-cobalt-copper-chromium, platinum group of minerals and gold rich anomaly that is 2km long and 250metres wide. Significant quantities of iron ore deposits were discovered in south-western Uganda in places such as Buhara, Muyebe and Nyamiringa in Kabale district; Nyamiyaga and Kazogo in Kisoro district; and at Kinamiro in Butogota Kanungu District. There are large deposits of phosphate at Sukulu in eastern Uganda.

Copper, tin, wolfram and iron have been exploited at different scales since colonial times, but their extraction declined or was halted altogether in the mid-1970s and 1980s as the world prices plummeted. Exploitation of these minerals is being revived as international demand and prices increase. Gold has been mined

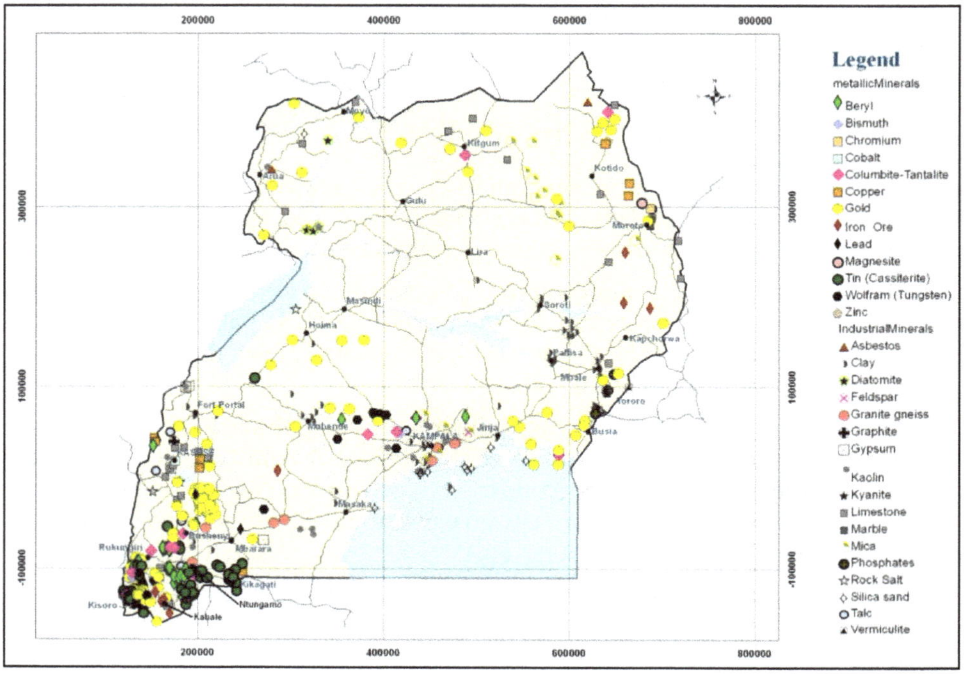

Figure 11.1: Mineral Occurrence in Uganda.
Source: Ministry of Energy and Mineral Development, Uganda

in Uganda at artisan levels. However, there is growing government and investor interest to increase investment in gold mining in Uganda. Similarly, there is growing investment interest in mining and exporting rare-earth-minerals. This, notwithstanding, the level of investment interest in these other minerals is not akin to that of fossil oil. A total of 867 licenses and certificates are currently operational.

In 2013, the country mined limestone, pozzolana, gold, vermiculite, cobalt, wolfram, aggregates, and kaolin iron ores worth Uganda shillings 208billion (US$81.5million). Of this, minerals worth Uganda shillings 69.9billion (US$27.4million) were exported that included copper, cobalt, gold, manganese, quartz, silver, tin, tungsten, and vermiculite.

Plans are underway to implement the Regional Certification Mechanism (RCM) of the Great Lakes region to minimize conflicts in the marketing of minerals from the region.

The mineral sub-sector earned the country a non-tax revenue accruing from royalties and mineral license fees totalling to Uganda shillings 13.8billion (US$5.4million).

Oil and Gas Prospects in Uganda

The country recently (1997-2006) discovered significant amounts (3.5billion barrels) of fossil oil reserves in the Albertine Graben in the western part of the country that the country (Figure 11.2). Between 1.2 and 1.7 billion barrels of these resources are considered recoverable with room for increase since only 40 per cent of the prospective area has been explored.

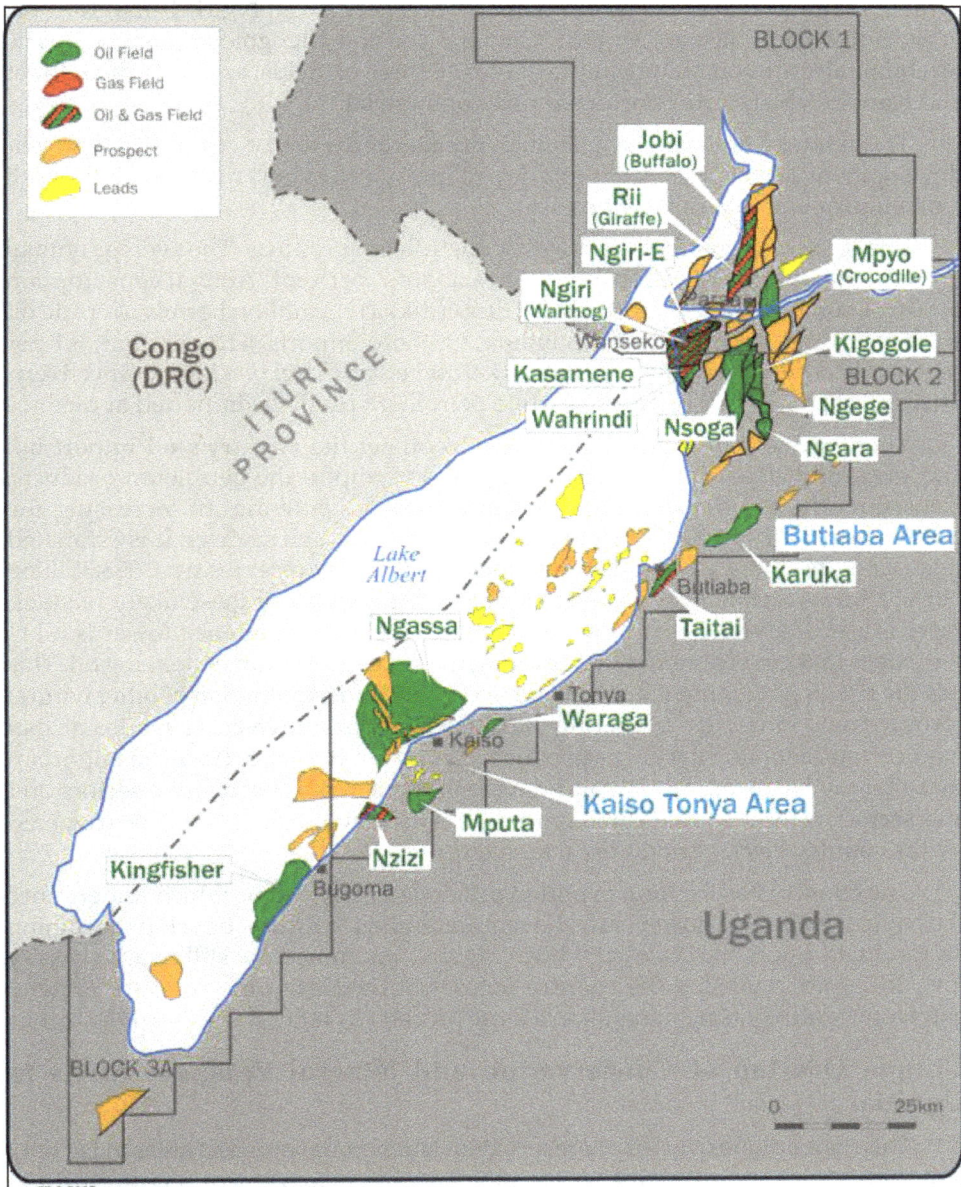

Figure 11.2: Uganda's Petroleum Prospects in the Albertine Graben

Source: Mr. Brian Glover, Tullow Uganda Operations General Manager, presentation at Logistic Suppliers Open Day, Uganda in January 11, 2011.

To date, a total of ninety exploration and appraisal wells have been drilled. Of these, seventy nine encountered oil and gas in the sub-surface, a success rate of 85 per cent which is among the highest success rates globally. Twenty one petroleum discoveries in the Albertine Graben *i.e.* Turaco, Mputa, Waraga, Nzizi, Jobi, Rii, Ngassa, Kajubirizi (Kingfisher), Kasamene, Ngege, Nsoga, Ngiri, Taitai, Karuka, Wahrindi, Ngara, Mpyo, Jobi-East, Gunya, Lyec and Kigogole (Figure 11.2). The oil companies are conducting appraisals of all these oil fields, except Turaco, Taitai and Karuka which were considered non-commercial.

The resource exploitation plan includes use of the gas or crude oil for power; development of a 60,000 barrels per day refinery; and export of crude oil through a pipeline or any other viable option.

The country currently imports all her petroleum products. The average annual growth rate of petroleum consumption is about 7 per cent. Annual consumption of petroleum products is 1,500 million litres (dikka12.3 million barrels of crude oil equivalent) which costs US$1.296 billion. Of these imports, 41.1 per cent, 6.1 per cent and 52.8 per cent are usually petrol, kerosene and diesel products, respectively. This excludes 1.303billion litres of white petroleum products imported annually.

The discovery of crude oil is likely to off-set the country's oil import bill and save funds that would otherwise be used to import the petroleum products. Consequently, the discovered oil resource has the potential of increasing the country's treasury reserves through oil sales and revenue savings. It is estimated that the oil resource alone is likely to increase treasury reserves by US$3.3billion per year once commercial production begins. As a result of these likely realities, there is a shift and increase in government's development interest towards oil to the extent that government is fast-tracking the process of commercializing oil. This has the risk of government ignoring development and exploitation of other natural resources and minerals – triggering the much touted resource-curse syndrome that has been common to oil and gas producing countries in Africa. It will be important that there is balanced growth and development across all economic sectors and sub-sectors, if the resource-curse syndrome is to be avoided. This will be triggered by the country having sound fiscal management.

The country has for long exported minerals in raw forms, which undermines the country's capacity to generate the required funds to further develop the mining sector. In addition, the scales at which minerals are mined are still relatively very low. It is important that the country focuses at increasing the scales of mineral extraction and exporting minerals in a more processed/refined/value-added forms.

Unique Overlap of Conservation and Mineral Prospect Areas in Uganda

There is a unique overlap of conservation and mineral prospect areas in Uganda that presents significant challenges to planning and management (Figure 11.3).

It is not clear how conservation areas coincided with areas that have significant deposits of minerals. Many of the conservation areas were identified and gazetted during colonial times. As a result, it is rumoured that Colonial Masters got to know

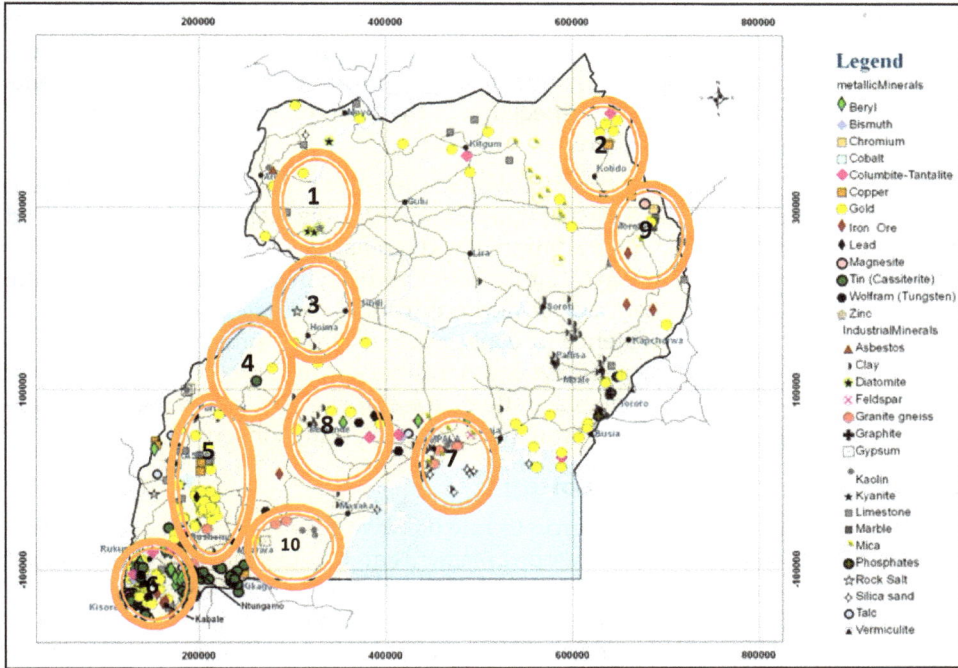

Figure 11.3: Unique Overlap of Conservation and Mineral Prospect Areas in Uganda.
Source: **Generated by Author**

Key to Overlap Areas in Figure 11.3

Overlap Area	Conservation Areas Overlapping with Mineral Areas in Uganda	
1.	– Lomunga Wildlife Reserve – Ajai Wildlife Reserve	– East Madi Wildlife Reserve
2.	– Kidepo National Park	
3.	– Karuma Wildlife Reserve – Murchison National Park – Kabwoya Wildlife Reserve – Budongo Forest Reserve	– Bugungu Wildlife Reserve – Ramsar Site (2006) – Kaiso-Tonya Wildlife Reserve
4.	– Rwenzori Mountain National Park	– Kyambura Wildlife Reserve
5.	– Queen Elizabeth National Park – Malamagambo Forest Reserve	– Toro-Semuliki Wildlife Reserve
6.	– Kigezi Wildlife Reserve	– Bwindi Impenetrable National Park
7.	– Mabira Forest Reserve	
8.	– Kibaale National Park	
9.	– Mount Elgon Forest Reserve	– Mount Elgon Wildlife Reserve
10.	– Lake Mburo National Park	

where vast minerals were located, so they gazetted the areas to protect them from human encroachment and for future exploitation purposes. Whatever the case, the overlap of protected conservation areas with significant mineral deposits presents huge development challenges, especially in the current legal dispensation that prohibits mining in protected areas.

Often, there is conflict between advocates for conservation and development planners on what constitutes the most judicious resource management option and best trade-off for the loss or degradation of a given natural resource. This is further complicated by the referred law that prohibits mining in protected areas and the usually weak or non-existent social and environmental safeguards that would otherwise ensure the co-existence of conservation and commercial development initiatives. This may require a review or amendment of the law on forests, water or wildlife resources and putting in place effective social and environmental safeguards to allow mining and other developments to co-exist with conservation.

Economic and Political Stability

Uganda's annual economic growth rate has averaged five percent (5 per cent) over the last 25 years. This is a rather promising trend that has the potential to attract internal and foreign direct investments. However, it is important to put this growth into perspective in regards to where the country is coming from economically and politically. After independence in 1962, the country enjoyed a period of relative economic strength and stability until the 1970s, when the country was ushered into a dictatorship under the leadership of President Idi Amin Dada. The country was also subjected to economic and trade embargos and armed conflicts during the 1970-1980 periods, which consequently undermined the economic growth situation and investment interest to zero. The 1980-1985 periods, although under a different political leadership, was marred with rebel insurgency, which did not improve the economic and political situation. The economic and political situation of the country begun to improve after the 1985 period with the ushering in of the current National Resistance Movement (NRM) regime that has ruled the country for over 25 years. As a result, the country registered 5.0-6.5 per cent annual economic growth rates, which by comparison, have been significant. However, these development achievements run a risk of being undermined as the country gradually shifts from a relative democracy to an autocratic political dispensation and corruption become a rampant vice.

A broad range of issues are emerging related to the extractive sector including environmental consideration to land rights and from transparency in the licensing process to management of incoming revenues. It will be important that the emerging issues are properly addressed, because if not properly addressed could be a recipe for conflict. It is also important that the current leadership recognizes the political developments and ensures democracy and economic and political stability, since the latter two are critical ingredients for attracting investment. In addition, it is important that there is genuine commitment to fight corruption, because corruption is a vice common, not only in the mining sector, but also other social and economic spheres. For example, corruption has become a cancer in the country and has crippled a number of government projects, making it impossible to effectively deliver services to citizens. Corruption, money laundering and tax evasion or avoidance are blamed for the significant (billions of dollars) illicit financial outflows from the country (Bazira *et.al*, 2012) and the African continent at large (Global Financial Integrity, 2012).

Mining Policy 2002 and Mining Act 2003

The vision of Uganda's Mining Policy 2002 is to attract investment, build capacity for acquisition and utilization of Geo-data and increase mineral production for economic and social development of Uganda. This is premised on the fact that minerals are non-renewable resources, which if not properly managed, their sustainability is jeopardized.

The policy sets out to provide conditions conducive to attract new investment for exploration and mining development, with the private sector providing the necessary management, technical and financial resources required.

Mining Policy envisaged a public trust doctrine where government protects minerals on behalf of the people of Uganda. A similar tenet is held in the 2003 Mining Act. However, with the amendment of the 1995 Constitution to include Article 244 that enshrines the custody of minerals and petroleum in "the Government on behalf of the Republic of Uganda" (*i.e.* the State), the amendment in a way took away the public trust doctrine in respect to minerals, but not the other natural resources such as gazetted forests, water bodies, wetlands, and wildlife areas [Article 237(b)]. The oil and gas policies and regulations espouse the tenets of the amended Constitution. This could lead to legal conflict in the development, use and management of minerals and other natural resources, especially where there is resource inter-linkage. Such legal conflict could easily be taken advantage of by unscrupulous investors to disenfranchise the country. This calls for harmonization of the policies and laws in respect to all natural resources management – including oil and gas.

Government of Uganda is currently in the process of revising its policies, laws and institutional frameworks to make them relevant to the demands, requirements and standards of the emerging oil and gas industry. It is important that the process of revising the policies and regulations does not get blinded and biased to oil and gas only, but to be comprehensive and reflective on all economic sectors and sub-sectors.

Oil and Gas Policies and Acts and Stage of Oil Development

Government of Uganda developed an Oil and Gas Policy in 2008. This was in response to promising oil and gas discoveries and emerging industry. The government also enacted petroleum laws in 2013 to govern the upstream and midstream petroleum value-chains. These petroleum laws replaced the 1985 Petroleum Exploration and Production Act which was found inadequate to fully take the industry from exploration into the production and commercialisation stages.

Government also put in place a Petroleum Revenue Management Policy 2013 to guide the use of revenues generated from petroleum extraction, refining and marketing processes for the development of the country and poverty reduction. This policy will be operationalized by an amended Public Finance and Management

Act. It will be important that the policies and regulations being put in place will useful and effective in containing illicit actions common among many extractive companies such as environmental abuse and illicit financial transfers.

At the time of preparing this paper, 40 per cent of the oil exploration area is licensed to multinational oil companies that are at different stages of development. The remaining 60 per cent of the Exploration Areas are still available for bidding. Only one company has been awarded a production license and is preparing to commence production. The others are still at exploration, field appraisal and development stages, but it is expected that they will gradually move into production and commercialisation.

Energy Resources for Industry Development

The existence of reliable and affordable energy is critical to the extractive industry. Uganda's energy for industry development originates from electricity mainly produced from hydropower stations. This is further supplemented by thermal power. About 450MW of installed capacity is available. However, this is not enough to meet the domestic, commercial and industrial demand. It is riddled with technical and commercial losses, which dwindles the actual amount of electricity available for distribution to about 50 per cent. Consequently, the available electricity is expensive and unreliable characterised by power-cuts and load-shedding. This is despite the fact that only 12 per cent of the households (including commercial or industrial units) in the country have access to this electricity. The cost of electricity is about US$0.23 (23US cents) per unit. This raises affordability and reliability concerns related to electricity supply. Such a cost easily undermines investor interest to invest in the country. The rising petroleum consumption and prices also worsens the energy security situation in the country.

Current peak energy demand is projected to be an average of 500MW. It is clear that demand outstrips the available installed power capacity. Government's broad objective to address the energy security issues is to provide adequate and reliable power supply to the country through increased generation capacity; demand-side management and the use of alternative sources of energy. Government's strategy to achieve its objectives is to enhance public-private partnerships in power generation and supply:

☆ To enhance financial sustainability of the power sector in light of the high energy costs arising out the escalating petroleum prices on the world market:

☆ To increase inter-regional power trade in East Africa;

☆ To address the declining Lake Victoria net basin supply and to arrest any environmental degradation

☆ To bring down the electricity cost to regionally competitive levels to attractive domestic and foreign investment.

☆ To exploit alternative and integrated energy options

Government is constructing another 450MW hydropower dam at Karuma on River Nile, which will be supplemented by additional mini-hydropower stations. Other planned hydropower stations include Ayago North (300MW), Ayago South (200MW), and Kalagala Falls (600MW). Construction of these additional power stations was slated for a 2012-2020 period. Even if this is achieved, by the time these power stations become operation, the available electricity will still be below the 2025 projected average peak demand of 1910MW, suggesting that additional energy resources will be required.

It is hoped that once oil and gas are commercialized, a portion of the oil and gas will be used in thermal power stations to add to the hydropower grid. This is likely to increase the amount of electricity available and lower the cost. Another important factor that government needs to focus on in lowering the cost of electricity is to increase the coverage (consumer numbers) *i.e.* economy of scale.

Human Resource Capital

Uganda has a population of about 35million people of which 75 per cent constitutes an active and productive youth. While the literacy levels of the country are still low, there is a growing body of literate youth with training in various academic field and skills, including oil, gas and mining. The growing literacy levels are attributed to the many emerging institutions of higher learning in the country. However, the youth will require more skills, experience and training in order to become more competitive with their foreign counterparts. This expertise gaps is making it a challenge for Ugandan youth to be employed in the extractive sector, compelling companies to employ foreign expatriates for jobs that would otherwise be done by Ugandans. This calls for more government investment in training Ugandan youth in a multiplicity of skills relevant to the private sector.

Proposed Actions for Ensuring Mineral and Mining Sector Benefication in Uganda

Mineral exploitation should be pursued to meet a country's broader social and economic goals, but not as an end in itself. This requires having a vision of how the resource fits in a country's economic future. For some countries, the best use could be to leave the resource untapped for a period of time, while for others, it may need to exploit the resource immediately to generate revenue and create wealth to sustain investments for growth and meet basic need of the citizens. Whatever the decision, there are practical guidelines that need to be undertaken to maximize the opportunities and minimize the challenges presented by any given mineral resource.

Key Guidelines for Mineral Resources Utilization in Uganda

There are eight broad key guidelines for mineral resource exploitation and they include the following:

☆ Mineral resource use for social and economic benefits;

☆ Government accountability to an informed public;

☆ Formulating fiscal and contractual terms that are robust to changing and uncertain circumstances;

☆ Competition in the award of contracts and development rights

☆ Careful decision-making whether or not to exploit taking into account local economic, environmental, social and political effects. Private sector led versus People centered development

☆ Clear, transparency and accountable revenue sharing, transfer and investment based on stipulated rules, procedures and sound judgment;

☆ Use natural resource revenues for additional wealth creation and social and economic growth and development;

☆ Caution in entering international trade and investment agreements/treaties.

This section briefly discusses the above guidelines in line with current government practice.

Mineral Resource use for Social and Economic Benefits

The development of a country's mineral resources should be designed to secure the greatest social and economic benefit for its entire people *i.e.* the current and future generations. This requires to formulating, implementing and monitoring detailed programs and policies in multiple areas, including leasing and fiscal regimes, social and environmental regulations and development plans. Decisions have to be made whether or not to take capital and capabilities from public or private sector for exploitation of the natural resource. In addition, decision making should be stepwise in the following order before projects are implemented

i) Evaluating the resource potential including but not limited to size of the mineral resource, location of the mineral, quantity of mineral reserves and ease of reach and exploitation of the mineral;

ii) Deciding whether or not to exploit;

iii) Formulating the fiscal, contractual and regulatory terms;

iv) Negotiating and endorsing contractual terms;

v) Establishing governance and oversight regimes;

vi) Agreeing on the processing and marketing regimes;

vii) Revenue management scheme

Uganda is excellent at formulating policies, regulations, programs of action and to some extent making decisions. However, she is weak at implementing or enforcing the policies, regulations and decisions. This is mainly attributed to a duplication of roles and responsibilities within and across ministries, departments and agencies and political influence peddling that undermines the institutions' ability to freely exercise their mandates. There is need to harmonize institutional roles and responsibilities and eliminate the duplication.

Government Accountability to an Informed Public

The 1995 Constitution of the Republic of Uganda gives citizens the right of access to information (Article 41). This is further augmented by the Access to Information Act 2005. However, the right of citizens' access to information is undermined by the existence of a 1963 Official Secrets Act that was not repealed upon enactment of the 2005 Access to Information Act and is further undermined by requirement for a Requester to pay a non-refundable fee representing the actual cost of retrieval and reproduction of the information. The fee easily disadvantages the poor from accessing public information. This is further worsened by mindsets within governments whether by commission or omission to believe that it is *"better to govern an ignorant population"*. This has the potential of bottling-up discontent and anger among the citizens and frequently causes citizen uprising, which is counterproductive. This mindset, although not expressly stated in Uganda, seems to be the underlying perspective based on common practice of denying information disclosure. It should be in government's interests to have an informed citizenry. It should also be in government's interests to avoid citizen uprising by providing proactive, accurate and timely information.

Generation of critical information for mineral resources exploitation should be proactively conducted by governments for citizens to use as basis for decision-making whether or not to invest in the sector. Feasibility studies for auxiliary industries utilizing up-, mid- and down-stream by-products of the mineral resources value-chain need to be strategically produced by governments as policy to encourage citizen investments in resource exploitation. Governments need to build sufficient information about a resource to better position them during contract negotiations *i.e.* purging the information disadvantage of government could be achieved through competitive bidding processes that usually avail the critical information. Government of Uganda has to a large extent endeavored to acquire, process and interpret geochemical and geophysical data of many minerals and has provided a website where such information can be obtained, which is a very good action. Government is also building human resource capacity in contract formulation and negotiation, but is still weak in this area. With the exception of the oil refinery, it has not been a tradition for government of Uganda to undertake feasibility studies for mining industry, because the studies are expensive, leaving the burden to private sector investors. However, it could be a useful intervention for government to invest in conducting feasibility studies for the mining industry - whose cost can be off-set from an investor and the study used as comparative reference scenario with investors' feasibility studies. Often, a country is lied to by unscrupulous investors, because of lack of corresponding or alternative augments to mining proposals.

Competition in the Award of Mining Contracts and Development Rights

There is a common practice of "sourcing mining bids directly" in the absence of application or competitors. While this is okay where it is genuine, in some places it is used as a ploy to bring companies of "choice". The practice of "sourcing directly" played-out when government of Uganda was seeking the initial investors in current licensed mining areas, including the oil Exploration Areas. However, Government has expressed commitment to applying competitive bidding in the next licensing rounds. Whether this will be followed through, remains to be seen. It is important that competition is enshrined in all mineral resource exploitation deals. This notwithstanding, it will be important politically to ring-fence certain sections of the mineral-value-chain in favour of local/indigenous investors to encourage them invest and help them overcome the challenges they face competing with already established companies. This, however, has to be done transparently.

Carefully Decide whether or not to Exploit

Questions should be asked why exploitation is being pursued. This should be based on evidence, but not political whims. The decision to exploit should be well thought on the impact of the decision on individuals, society in general, economics, environment and politics at local community, national and trans-boundary levels. The decision should be people-centred not private-sector centred. The argument that private sector-led development is the best path is no longer tenable.

In Uganda, there is a semblance of people-centred development. This is depicted in the mining and petroleum laws that offer mining royalties and dedicated budgetary allocations to communities leaving within and in proximity to mining areas, but whether the communities have actually been receiving the payments and the budgetary allocations have achieved (are achieving) the desired societal beneficiation, is not very clear. The reality on the ground is contrary.

Transparency and Accountability

Transparency should be the rule, not the exception. There are too many confidentiality clauses in contracts, especially those dealing with mineral resources. Confidentiality clauses where a public natural resource is concerned should be restricted to patent, proprietary and privacy rights and security risks, which must be ratified by parliament or law. Otherwise everything might be classified. The whole process of negotiation and contract award should be transparent. A rules-based and transparent movement or transfer or investment of revenues based on sound judgment must be followed. Controversy surrounded the disclosure of Production Sharing Agreement (PSAs) signed between government of Uganda and oil companies to the extent that citizens petition courts on the matter, which yet to be resolved.

There is evidence that natural (mineral) resources management is more successful where there is transparency and an informed public. It requires government to:

☆ Posses political will, capability and capacity to take difficult and complex decisions and implement them; and continuously and proactively update citizens about natural resource exploitation and use. This is a tenet enshrined in the Universal Declaration of Human Rights; the Rio Declaration; the Aarhus Convention; the OECD Guidelines; and the IMF Code of Good Practice on Fiscal Transparency;

✰ Accept public scrutiny and demand for decision-makers to be held to account

Transparency along the entire mineral resource value-chain is critical. Any information to be withheld from the public must be proscribed and justified by Parliament. There is limited evidence that Uganda is committed to this principle and practice.

Use Mineral Resource Revenues for Additional Wealth Creation and Social and Economic Growth and Development

Mineral resources usually generate a lot of revenues, but dependence on mineral revenues is limited by the fact that the resources are finite and exhaustible and the revenues volatile and unpredictable. Therefore, in order to have sustainable benefits from mineral resources, the revenues must be utilized to put in place systems and infrastructure that ensure continued socio-economic benefits. For example, Future Generations deductions could be made on the mineral revenues and invested in low risk ventures to create additional wealth and cater for future revenue needs.

The decision to spend mineral revenues will be guided by the social and economic situation and development aspirations of the country. However, thrift spending of mineral revenues could trigger inflation, the Dutch disease, corruption, damage other industries/economic sectors, political conflict, widen the gap between the rich and poor and worsen the poverty situation of the country.

In addition, mineral revenue management schemes need to be designed to address the following issues:

✰ Address revenue volatility;

✰ Smoothen public expenditure to limit consumption expenditure

✰ Promote investment of revenues for growth and development, including poverty alleviation, infrastructure, health and education; availing low interest and long-term financing through the banking systems to indigenous private sector businesses

✰ Promote accountability;

✰ Enforce a rules-based and transparent revenue sharing/transfer scheme;

✰ Take into account trade-offs related to value-addition processing and diversification of the economy

In Uganda, management of mineral revenues has essential been managed at the central government level, with limited involvement of the local governments where the minerals are mined. The mineral revenues are collected by the Ministry of Energy and Mineral Development (MEMD) and the Uganda Revenue Authority (URA) and deposited into the national treasury in the Bank of Uganda. They become part and parcel of the overall treasury pool. There is no separate dedicated Minerals Fund into which mineral revenues can be deposited and monitored on how they are utilized. This is because the mineral revenues are not generated rapidly and are comparatively small to justify a special Fund. However, with continued expansion of the mineral sub-sector and the emerging of the petroleum sub-sector that is characterized by rapidly generated and large sums of revenues, it will be important that a Special Mineral Fund is created to monitor the revenue transfers. Government of Uganda has recognized the need for a special fund and is planning to establish a petroleum fund into which petroleum revenues shall initially be deposited before they are transferred to the other relevant accounts. While it is not advisable to have a multiplicity of special funds, because of the challenges/difficulty of managing them, it might be important to have one consolidated special funds

Caution in Entering International Trade and Investment Agreements/ Treaties

Caution needs to be taken when entering into international trade and investment agreements. It is important to understand what a country is likely to lose or gain before signing the agreement. Many trade and investment agreements are drafted by others and skewed in their interests.

Governments need to be given time to scrutinized international trade and investment agreements, with the involvement of the public or civil society, before they are signed to avoid being taken advantage of.

Government of Uganda has signed many trade agreements/treaties that are rather lope-sided in favour of the other party. This makes it rather difficult for Ugandan-based firms to operate abroad in jurisdictions that have an agreement/ treaty with Uganda. Independent effort is currently going on to review Uganda's international trade and investment agreements to assess the pros and cons of the agreements/treaties in respect to the country, with a view of seeking amendments.

Establishing Governance and Oversight Regimes

It is important that governments put in place policy, legal and institutional frameworks to govern respective natural resources; strengthening oversight and transparency rules for accountability; and establish independent auditors/ resource/fund managers and public interest accountability entities to support the Parliamentary oversight.

As earlier mentioned, government of Uganda is reviewing her policy and regulatory framework to meet the demands, challenges, requirements and standards of the emerging oil and gas industry. It is hoped that the amendments will also effectively apply to the mineral industry. It will be important to benchmark the new policies and regulatory frameworks to the mining sector.

Traditionally, Parliament is the main oversight body. However, it is increasingly becoming clear that many Parliaments alone usually lack the time and requisite human and technical capacity to synthesize, analyze and provide oversight on key mining concepts and documents, suggesting that they would require external support. It is for this reason that governments need to strengthen the oversight and transparency rules by allowing external support to parliament. This may be achieved by the creation of independent bodies that support the work of parliament and other regulatory bodies, for example, establishing Public Interest Accountability Committees (PIAC), Independent external auditors, independent supervisors, *etc.*

Civil Society Engagement for Oversight Enhancement;

The role of civil society is critical in promoting transparency and accountability and oversight. Governments must allow civil society and the media (print and electronic) to operate freely and without harassment and intimidation. It is common practice for government to appoint civil society representatives for multi-stakeholder discussions, this should stop. Civil society should be free to nominate their representatives independently. A

Enforceable and Prohibitive Conditions and Penalties

Often, the conditions and penalties accorded to breach of contract or responsibility at government level are weak and not prohibitive in nature. Consequently, they are abused/violated. For most mineral resource exploitation companies, the penalties are equivalent to "loose pocket change" that they would prefer to pay and continue violating the conditions, especially if it is cheaper to violate than to meet the demands of the conditions. The current conditionalities and penalties for mineral/natural resource extraction in Uganda are weak and not prohibitive at all.

It is important that governments place encumbrance and prohibitions to collateralization of natural (mineral) resources.

Conclusions

☆ Uganda is not starting from scratch. There is already existing knowledge and expertise internationally that could be taken advantage of;

☆ Work towards putting in place policy, legal and institutional frameworks to ensure sustainable petroleum resources use has already begun, but more is needed to make the policy and regulatory frameworks comprehensive and cover also the mineral resources;

☆ There are vast quantities and fair distribution of mineral resources, including petroleum in Uganda that are still underdeveloped and underexploited and provide an opportunity for investment;

☆ There is already an established international market for the minerals found in Uganda;

☆ Mineral resources are finite, therefore their judicious exploitation is critical for Uganda's growth and development and socio-economic sustainability;

☆ East Africa Community (EAC) countries have an Energy Master Plan to harness the energy resources in the region. Therefore, they could develop other master plans related to natural/mineral resources and they can muscle the necessary capital for investment to harness the available natural resources, if they genuinely choose to work jointly.

Recommendations

☆ Uganda needs to promote her mineral resource potential internationally to attract investment;

☆ Government needs to undertake mineral feasibility studies that they could sell to potential mining companies inside and outside the country and also use them as reference scenarios for similar independent analyses;

☆ There is need to provide low interest and long-term financial services to indigenous companies to encourage them invest in the different sections of the mineral-value-chains as a means of expanding investor participation in the mineral sector;

☆ Government should insist on prior mineral processing and value-addition locally, before export;

☆ Government needs to inculcate the culture of competitive bidding in the extractive sector, even when there is ring-fencing of certain sections of the mineral-value-chains;

☆ There is need to balance the conservation and development interests;

☆ Government needs to safeguard the economic and political gains so far achieved from sliding back into anarchy;

☆ There is need to build the human, financial and technological capacity of Ugandans in dealing with the mining sector;

☆ There is need to strengthen the oversight and accountability frameworks in the country;

☆ There is need to use mineral revenues for social and economic transformation and sustainability;

☆ A careful process needs to be undertaking when making decisions on whether and how to exploit a given mineral resource.

☆ A framework on how Uganda can work with her East African Community neighbours to exploit the mineral resources needs to developed and agreed upon.

Chapter 12

Cost Benefit Analysis of Bauxite Exploitation in Tay Nguyen Area, Vietnam: Lessons for industrial Mining in Developing Countries?

Vu Anh Thu

Department of Geology,
Hanoi University of Mining-Geology,
Ward Dong Ngac, Tu Liem District, Hanoi, Vietnam
E-mail: vuanhthuvn@gmail.com

ABSTRACT

Bauxite exploitation has created various impacts in the regions of the Tay Nguyen, Vietnam. However, it is not clear how to determine between cost and benefit from bauxite exploitation projects in this region. Mining industry may be the dominant industry in the local community and in the region in terms of providing local employment and generating income; it does not necessarily have direct linkages to the local economy and therefore does not contribute fully to diversified sustainable development of the local community or the region. This paper discusses the cost and benefit of bauxite exploitation projects in the Tay Nguyen area. The regional economic diversity is identified using the input–output analysis. This paper attempts to identify the key sectors, backward and forward linkages in the Tay Nguyen area in order to analyze which impact in the region are needed to be considered to increase their connections with the mining industry to induce higher retention of benefits from industrial mining.

Keywords: Bauxite, Cost-benefit analysis, Environment, Exploitation, Industrial mining.

Introduction

Located in the Central Vietnam, Tay Nguyen area has various resources and long time culture of ethnic minorities. The area is around 54.641.0 km², with approximately 1.8 million ha of old-growth forest and various ecosystems, the covered ratio of forest size is about 32 per cent (Nguyen, 2013). With a 500 -1500m of high from sea level, this area known as "a roof of Indochina" which has a potential basalt rock with a thickness of about 500-600m. According to previous studies, Tay Nguyen area has a potential of bauxite ore with the reserve is around 5.5 billion tons (Dang, 2009) The population in this region is 4,059,928 with the density of 93 people/km² and the population growth is 5 per cent. In addition, the Tay Nguyen is a unique area in Vietnam with majority of ethnic minorities with poor livelihood and education (Nguyen, 2013).

Since 2013, the Vietnamese government has carried out 2 projects about bauxite ore in the Tay Nguyen area (from 2012 to 2025) to convert in to alumina for the production of aluminium metal. The capacity has estimated around 5-6 million ton of ore/year. These projects have opened extremely debate not only in Vietnam, but also in some countries. Some potential risks of these project have questioned for governors, scientists as well as economists. These projects lacks of social impact assessments and environmental impact assessments. Important criteria (such as: the electric power and water available, location, cost of transport, reserve of ore, human resource) to decide to carry out bauxite ore as well as potential hazards did not mentioned. In addition, environment issues in the Tay Nguyen will become polluted and destroyed when thousand ha square of old-growth forest will be deforested and a thousand tons of red solid waste still not do not have any optimal method to treat.

In terms of society, ethnic minorities in the Tay Nguyen have long and special culture. The UNESCO recognized Space Cong Culture in the Tay Nguyen as a Masterpiece of the Intangible Heritage of Humanity on November 2005 (Nguyen, 2013). Their livelihood is based on the cultivation of industrial trees and forest. Immigration issue to run this project, local people have to face difficult situations as these projects, lacks of sustainable development for ethnic minorities to adapt to new work. With the change of traditional habit, it means the Space Cong Culture in the Tay Nguyen also fall into oblivion.

This paper will focus to clarify the cost-benefit analysis about the two projects of bauxite ore exploitation in Tay Nguyen area, Vietnam toward the sustainable development for the region.

Materials and Methods

This study develops a rigorous and innovative methodology to measure, model and value the impacts of mining on the physical and social environment. The approach followed was to design a model based on Excel, using data generated from the mining sector. The Excel-based model model consists of linked spreadsheets. The basic operating module consists of four spreadsheets:

☆ Benefits to the Mine

☆ Costs to the Mine

☆ Benefits to the Environment

☆ Costs to the Environment

Here, environment includes both the physical and social parameters.

Figure 12.1: The Location of Tay Nguyen Area in Vietnam

Mine Data

Mine benefits are clearly defined. They are represented by a profit margin to the mine operators and the payment of taxes, tariffs, fees and royalties to the government. Although mine operators were reluctant to disclose confidential financial information on their profitability. The percentage of royalties to be paid is laid down by law and therefore not difficult to calculate.

Mine costs, in terms of operating costs, are readily available; in fact, sometimes too much information is forthcoming. Mines do include environmental rehabilitation costs in their budgets and "donations" to local communities.

Environmental Data

Environmental benefits from mining are virtually non-existent, as the maximum mine operators are required to do under rehabilitation is returning the mined area to its original state. This does not confer any benefits. Payment for use of natural resources can be construed as a benefit for the environment not a benefit for the environment, but it is arguable if a true value is being assigned to such usage. For example, the water regulations stipulate tariffs, but allow users to calculate their own usage and fee. This encourages under-reporting. Environmental damage has been fairly well documented in Vietnam. Ministry of Environmental and Natural Resource (MENR) has compiled national figures for loss of natural resources and pollution. At the same time, the assigning of a cost figure to this damage is still in a very preliminary stage, although charges for use of natural resources are well documented in existing regulations. MENR have developed a somewhat complex methodology for calculating the costs of environmental damage which are based on empirical formulae in a Vietnamese published textbook. At the moment the valuation technique is still theoretical and could be challenged in the courts by the mines. A crosscheck on the values calculated would be useful and a pilot study of actual values of lost grazing lands is proposed under this study.

Socioeconomic Data: Socioeconomic benefits from mining are mainly reported in financial terms and there is a large amount of data on monetary flows in human development funds and sovereign wealth funds. The means by which this translates into actual social benefits for local people is less well understood and the whole issue of investment of government mining revenues is under review. Social impacts of mining are reasonably well understood. At a macro level, effects are recognized as a potential threat and at the individual level, impacts such as a higher divorce rate due to separation of workers' families is suspected, but little attempt has been made so far in Tay Nguyen area, Vietnam to assign real costs to these impacts. A pilot study of actual divorce rates and emotional stress is proposed under this study. An important point is that the usefulness of the model will be limited by the quality and accuracy of the data entered, and that will depend on full cooperation from the mining sector and from the government, to find the correct balance between the mine operations and the physical and social environment.

Cost Benefit Analysis

It seems the bauxite exploitation project in the Tay Nguyen area have a lack of sustainable development in the region. Based on previous studies about aluminum industry in the World, the scientists and economists conclude that it should be exploited bauxite if it satisfies 6 criteria in order: 1. Available power resource; 2. Available water resource; 3. Suitable location; 4. Lowest cost for transportation; 5. Potential reserve and 6. Cheap human resource. With the capacity of about 1.2 million of tons per year and will increase 6- 8 tons/year in 2020-2025, However, in Tay Nguyen area only satisfies the last criteria that is potential reserve and cheap human resource. As a consequence, the price of aluminum will be high and very costly to treat the environmental pollution issue in this region.

In addition, there is a lack of careful analysis carefully about produced ore or processed ore. If produce processed ore-aluminum, it means the price of exporting of aluminum is high, however the cost of environmental and hazard are potential and insufficient power available and treatment for red muddy waste to produce aluminum. In case, production of ore seems not optimal, option is to consider the

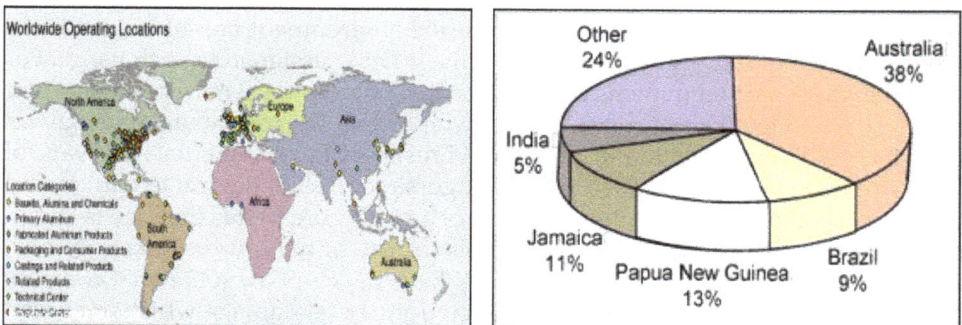

Figure 12.2: The Location and Aluminum Reserve in the World

current market situation according the world economic crisis. There are some issues are needed, consider in terms of *"cost-benefit analysis"*:

Transportation

This is an important issue in each exploitation project, particularly in Tay Nguyen area which is a mountainous and highland region. The plan to transport and carry ore and materials have not considered carefully. Because it need various transporters to carry different materials and ore or aluminum by two ways. It means the return way the transporter is empty with the distance is about 300 km. It is extremely costly. In addition, until now there are four plans to transport ore after producing is: by waterway, land, railway or pipe. It seems none of them is optimal option for the region. The first option is transported by railway, the cost to build the system is around USD 3 billions and will take until 2010 to finish construction (Dang, 2009). The land and waterway options are impossible in a highland area with high slope and long distance and cause environmental issue in the coastal zone area and negative impacts for developing tourism in the region.

Figure 12.3: Plans for Transport of Processed Ore in the Tay Nguyen Projects

Technology

According to the investor-VINACOMIN (Vietnam national coal-mineral industries holding corporation limited) of this project, the technology is applied by the CHALIECO company (China) under EPC format (Enginering - Procurement – Construction). It seems this is the cheapest technology to exploit bauxite in the region. This method has usually been applied to exploit mono-hydrate (Diaspore ore) in China, however the type of ore in Tay Nguyen is tripper hydrate (Gipsite ore). The selection of technology is also having problem with this project. As used to EPC, there are around 1500 Chinese workers will undertake to install the technology, it means local people in this area will miss opportunities to work on the projects (Hao, *et al.*, 2010, Liu *et al.*, 2009).

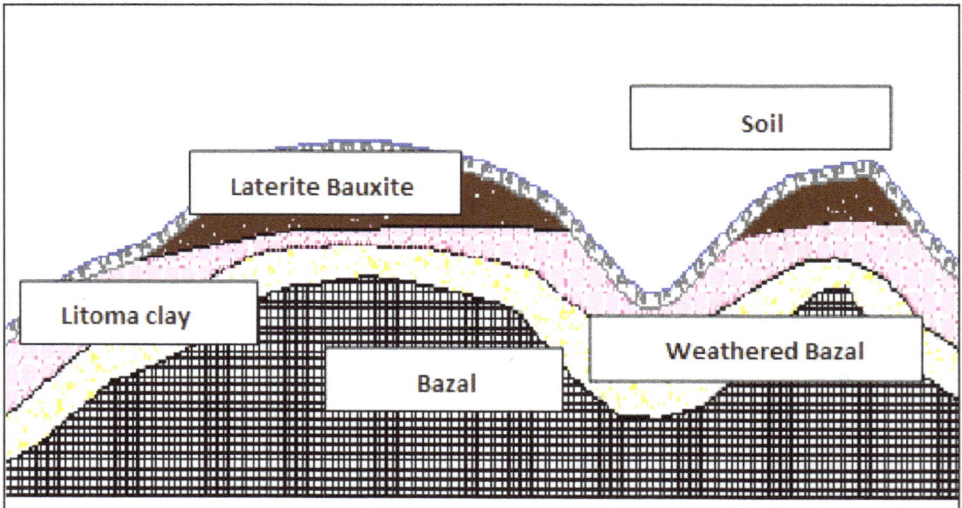

Figure 12.4: The Particular Geological Succesion in the Tay Nguyen Area

Environmental Issues

There are some environmental concerns arising from these projects. One of the most serious environmental issues is a red muddy waste after processing ore. Until

Figure 12.5: The Factory of Aluminum Ore Process in the Tay Nguyen Area

now, there is no potential technology to treat this kind of waste. The Tay Nguyen is a highland area with upper streams and has 6 months of the dry season. The red mud waste can damage lower streams in the future (Dai *et al.*, 2012, Anand *et al.*, 1996).

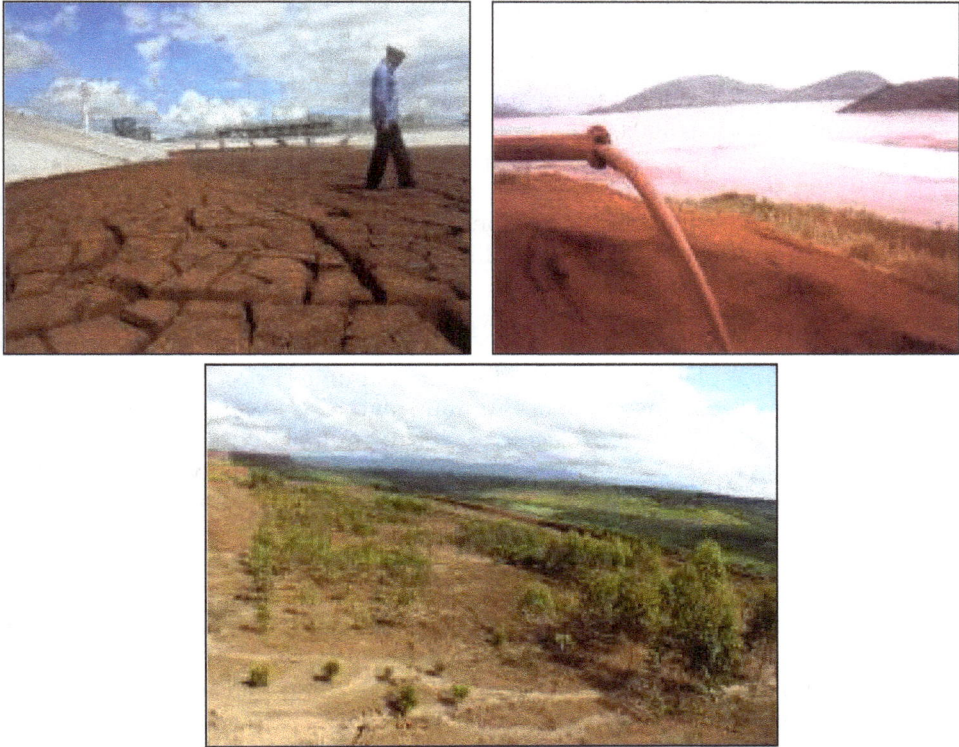

Figure 12.6: Environmental Issues in Tay Nguyen Area

Rehabilitation

To carry out the projects, there are a thousand hectares of growth-forest and tress industries will be deforested. It takes a long time to rehabilitate and the investor do not mention the potential method. Based on experience from Australian and The United States, normally to rehabilitate after bauxite exploitation, it needs about 20 years to cultivate for material trees not for farming (Courtney, 2013, Ivanova, 2014). In addition, the reserve of bauxite ore in the Tay Nguyen region is about 5.5 billion

Figure 12.7: Environmental Issues in the Tay Nguyen Area

tons, but the content only arranges from low to average (0.4-0.5) and contribute under spot with thin layers. It means the covered soil will be removed 3-4 times of thickness. As a consequence, the rehabilitation equals 3-4 times compare to Australia or Brazil with the same character.

Electric Power

In electrolysis, to collect 1 ton of aluminum ore it need 14.500- 15.500 kWh. To carry out electrolysis of 100.000 tons of aluminum/year, it needs a hydroelectric power with a capacity of 300-350 MW and the cost of investment for the station is about $4500-5500 USD/ton of capacity (Samal *et al.*, 2013). In addition, Vietnam consumes around 100-150 thousand of tons of aluminum per year, and the markets in the World produces about 70-90 million of tons of this. Thus, the market is sufficient, Vietnam can import easily.

Water

Water resource plays an important role in aluminum process. To produce a ton of aluminum needs $30m^3/1$ ton of aluminum and recycle water is about $12m^3/1$ ton of aluminum. The plan of project to exploit 1.2 million tons of aluminum/per year, it means that needs 24 million m^3 of water/year. The capacity from 2020-2025 estimates around 6-8 million tons of aluminum/per year, it is impossible to catch a demand for 160 million m^3/year for these projects.

Social Impacts

The Tay Nguyen is located major ethnic minorities with poor education and low living standard. The livelihood of local people based mainly on cultivate industrial trees and forest and they live in a huge Rong house for generations. With this particular custom of local people here, the UNESCO recognized Space Cong Culture in the Tay Nguyen as a Masterpiece of the Intangible Heritage of Humanity on November 2005. To carry out the project, there is an immigration of local people, they have to move into new areas and live in houses with cement material, not a Rong house that made from wood. Moreover, with a poor education, it is very difficult for local people can find a suitable job in these projects. It means local people may not get benefit from the project.

Figure 12.8: Rong House and Cement House of Ethnic Minorities before Project and after Emigrant

Figure 12.9: Traditional Cultivation (The rubber and coffee tresses) of Ethnic Minorities

Figure 12.10: Primitive Forest has Damaged in the Tay Nguyen Region

Conclusions

It seems the Tay Nguyen projects seems to have some problems and lack of cost benefit analysis. It can cause many serious issues for the region and difficult for sustainable development of local people and the region. The main issues need consider for carrying out the project such as: available water resource, available power, environment, lowest cost for transportation; potential reserve and cheap human resource. These projects bring back small benefit for the region. However, the consequence of exploiting activities were impacts on society, environment and nature in Tay Nguyen area. These projects need more research in the near future to ensure sustainable development to guarantee not only develop economically, but also to protect the environment, forest and livelihood as well as traditional customs of local people.

References

1. Dauvin, J.C., 2010, Towards an impact assessment of bauxite red mud waste on the knowledge of the structure and functions of bathyal ecosystems: The example of the Cassidaigne canyon (north-western Mediterranean Sea), Marine Pollution Bulletin, vol. 60, pp.197–206.

2. Samal., Ray, A.K., Bandopadhyay., 2103, Proposal for resources, utilization and processes of red mud in India — A review, International Journal of Mineral Processing, vol. 118, pp. 43–55.

3. Dai, S., Ren, D., Chou, C.L., Finkelman, R.B., Seredin, V.V., Zhou, Y., 2012, Geochemistry of trace elements in Chinese coals: A review of abundances, genetic types, impacts on human health, and industrial utilization, International Journal of Coal Geology, vol. 94, pp.3–2.

4. Ivanova, G., 2014, The mining industry in Queensland, Australia: Some regional development issue, Resources Policy, vol. 39, pp. 101–114.

5. Nguyen, T., 2013, Negative impacts of bauxite projects in the Tay Nguyen area, Vietnam, viewed online on June, 2014. http://www.hoahocngaynay.com/vi/nghien-cuu-giang-day/bai-nghien-cuu/214-tu-bauxite-den-nhom.html

6. Dang, D.C., 2009, Bauxite project in Tay Nguyen area, Vietnam, viewed online on June, 2014, http://vietsciences.free.fr/vietnam/donggopxaydung/khaithacbauxitetaynguyen02.htm

7. Anand,P.,Jayant M. Modak, J.M., Natarajan, K.A., 1996, Biobeneficiation of bauxite using Bacillus polymyxa: calcium and iron removal, International Journal of Mineral Processing, Vol. 48, I.1–2, pp. 51–60.

8. Hao, X., Leung, K., Wang, R., Sun, W., Li,Y., 2010, The geomicrobiology of bauxite deposits, Geoscience Frontiers, Vol. 1, I. 1, pp. 81–89.

9. Liuhttp://www.sciencedirect.com/science/article/pii/S0301751609001823, W., Yang, J., Xiao, B., Review on treatment and utilization of bauxite residues in China, 2009, International Journal of Mineral Processing, Vol. 93, I. 3–4, 8, pp. 220–231.

10. Lü, Q, Dong, X., Zhu, Z., Dong, Y., 2014, Environment-oriented low-cost porous mullite ceramic membrane supports fabricated from coal gangue and bauxite, Journal of Hazardous Materials, Vol. 273, pp. 136–145.

11. Courtney, R, T., Harrington, T., Byrne, K.A.,. 2013, Indicators of soil formation in restored bauxite residues, Ecological Engineering, Vol. 58, pp. 63–68.

Chapter 13

The Potential of using Natural Malaysian Silica Sand to Produce Leucite Glass-Ceramics suitable for Restorative Dental Applications

Malek Selamat and Siti Mazatul Azwa Saiyed Nurddin

Mineral Research Centre,
Mineral and Geosciences Department Malaysia,
Jalan Sultan Azlan Shah, 31400 Ipoh, Perak, Malaysia
E-mail: malek@jmg.gov.my

ABSTRACT

The aim of this research work was to investigate the effect of using natural Malaysian silica sand as the SiO_2 raw material on the crystallization, mechanical and biological properties of leucite (KAl_2SiO_6) glass-ceramics. Glass in the system of SiO_2-Al_2O_3-K_2O was prepared by melting the raw materials, quenched in deionised water and dry milled to obtain glass powder. The powder was pressed and sintered to obtain glass-ceramics. The thermal analysis, phase composition, microstructure, flexural strength, in-vitro bioactivity and cytotoxicity of the glass-ceramics were investigated. Thermal analysis defined the crystallization of glass in the range of 650°C and 850°C. The crystallization depends on the temperature and time duration of sintering. At 700°C, leucite began to form with minor phase of sanidine. The peak intensity enhanced with increasing the temperature up to 850°C. For sintering duration between 3 to 12 hours, the peak intensities of leucite and sanidine increased but microcline was formed as a minor phase. The microstructure analysis shows dendritic leucite and prismatic sanidine. The surface developed a glassy texture with no obvious micro cracks.

The leucite glass-ceramics appeared translucent. The flexural strength values (80–175MPa) were comparable presence of with that of a commercial product (112-140 MPa). The in-vitro bioactivity results prove that the leucite glass-ceramics sample can be classified as a bio-inert and non-cytotoxic material and can be used for restorative dental applications.

Keywords: Silica, Leucite, Sanidine, Glass-ceramics, Dental, Bioactivity, Bio-inert, Cytotoxicity.

Introduction

When the global tin crisis occurred in 1985, the Malaysian Government identified that industrial mineral resources such as silica sand, limestone, clay and feldspar had a potential for mineral diversification and expansion. There are widely distributed primary and secondary sources of these minerals in the country. Malaysia has high quality silica sand resources that exist in natural forms with reserves of about 148 million tonnes. Malaysia also has about 72 million tonnes of silica sand reserve from tin ex-mining tailing sand, which also contain heavy minerals such as ilmenite, hematite, zircon and cassiterite. In Malaysia, most of the silica sand producers use physical processing such as washing, screening, scrubbing and separation to remove the heavy mineral. The silica sand resources are presently used for low grade products such as in glass, ceramics and construction industry (Ahmad, 2006; JMG, 2012). Therefore, research studies have been done by Mineral Research Centre to add value and find better utilization of the Malaysian silica sand, especially for the development of advanced materials.

Silica sand is the main raw material for production of silicates glass. In general, glasses are not advanced materials but through refining, heat treatment and manufacturing, the glasses can be transformed into a new class of material called glass-ceramics. Glass-ceramics are polycrystalline materials and produced through controlled crystallization during heat treatment process of glass. Currently, the application of glass-ceramics as biomaterials for dental restoration is growing. For example, leucite glass-ceramics are widely used in dentistry as restorative dental material to fabricate dental inlays, crowns, bridges and veeners prostheses. Leucite glass-ceramics can be produced by sintering with surface crystallization of glass powder. This processing route involves melting, quenching, milling of glass frit, and sintering in order to promote crystallization of glass-ceramics. Glass-ceramics based on leucite show exceptional biocompatibility, and good physical, chemical and mechanical properties (Cattel *et al.*, 2005; El-Meliegy and Noort, 2011). The aim of this project was to investigate the effect of using natural Malaysian silica sand as the SiO_2 raw material on the phase crystallization, microstructure, flexural strength, in-vitro bioactive and cytotoxicity of leucite glass-ceramics.

Materials and Methods

Natural silica sand sample was taken from the Kampung Kolam deposit in Terengganu. The sample was wet screened and a fraction 75< to <150 µm was prepared in order to comply with the requirement of traditional glass production technology for glass melts. Chemical composition of the silica sand sample was determined by X-ray fluorescence analysis (Rigaku, Japan). A base glass composition of 64.2 per cent SiO_2, 16.1 per cent Al_2O_3, 11.9 per cent K_2O, 5.1 per cent Na_2O, 1.7 per cent CaO, 0.5 per cent TiO_2 and 0.5 per cent LiO_2 by weight was mixed in a milling jar for 5 hours and melted in a platinum crucible in an electric furnace at 1500°C for 2.0 hours. The melt was quenched into deionized water to produce glass frits. Then, it was dried and milled in a planetary ball mill (PM400, Retsch, Germany) for 5.0 hours to obtain glass powder, and later was screened to the required size of less than 75 μm. The crystallization temperature of the glass powder was studied using Differential Thermal Analysis, DTA (Linseis, Germany) with a 10°C/min heating rate at temperature from 25°C to 1100°C in a dry air atmosphere. The glass powder was cold pressed using a laboratory hydraulic hand press (Carver, USA) to obtain green compact in the form of cylindrical ingots of 13 mm diameter. The cylindrical compacts were heat treated in electric furnace (Termo Temp, UK) at 700°C, 750°C, 800°C and 850°C at a heating rate of 2°C/with and 1.0 hour soaking time to study the crystallization behaviour of glass. The effect of sintering duration on crystallization was also by through soaking for 3.0, 6.0, 9.0 and 12.0 hours.

X-ray Diffraction, XRD (D8 Advanced, Bruker, Germany) technique was used to identify phases present in the sintered glass-ceramics samples. Samples were analysed at room temperature using Cu Kα radiation at a scan speed of 2°/minute for 2θ from 10° to 80°. Field Emission Scanning Electron Microscopy, FESEM (Supra 40VP, Germany) was used to examine the microstructure of the glass-ceramics produced by various heat treatments. Strength is an important criterion having major influence on the clinical success of dental restoration. The flexural strength of sintered glass-ceramics was measured using three points bending test with bars of 20mm x 5mm x 5mm (INSTRON, UK, 0.5 mm/min displacement).

The biocompatibility of glass-ceramics sample after heat treatment at 850°C for 9.0 hour was examined by in-vitro bioactivity and cytotoxicity test. The in-vitro bioactivity test was conducted in Kokubo's simulated body fluid, SBF, which contains almost the same inorganic constituent as human body plasma. The sample was brought into contact with SBF fluid for 10 and 20 days. XRD was used to characterize the formation of apatite on the glass-ceramic surfaces. Cytotoxicity of the glass-ceramics material was evaluated by testing according to ISO 10993-5: 2009 (ISO: 2009). The tests were performed on extracts prepared by elution of the test samples in Dulbecco's modified Eagle's medium (DMEM) supplemented with 10 per cent foetal calf serum (FCS), at 37°C for 24 hours. China hamster lung L-929 cell was used to evaluate the cytotoxicity of the glass-ceramics via the 3-(4, 5-dimethylthiazol-2-yl)-2 5-diphenyl tetra-zolium bromide (MTT) assay. The cells were treated with the concentrations of 6.25, 12.5, 25, 50 and 200mg/ml of the test material for 24 hours. The cytotoxicity was determined by assessing the cell viability through the reduction of MTT. Cell viability was obtained by dividing the mean optical density (OD) values of the test material with mean OD of negative control and multiplied by 100.

Results and Discussion

The chemical analysis results given in Table 13.1 show that the natural silica sand sample is highly pure. The content of Fe_2O_3 is in the allowable range of 0.02 –

0.03 per cent which is suitable for glass production (Malaysia Standard MS 701:1981). The XRD diffraction pattern as shown in Figure 1 identifies that main mineralogical content in the silica sand sample is quartz. Quenching the glass melts in water at room temperature resulted in transparent and colourless glass frit. The XRD diffraction pattern of the glass sample as shown in Figure 13.1 indicates amorphous glass phase and did not show any evidence of un-molten batch materials in the glass. The DTA result of as-quenched glass powder with a 10°C/min heating rate at 25°C to 1100°C is shown in Figure 13.2. The DTA was used to identify the glass transition and crystallization temperature of the glass sample. The amount of heat released or absorbed as a function of temperature of a glass can give key information for glass-ceramics fabrication. The glass transition temperature of SiO_2-Al_2O_3-K_2O glass system is about 594°C to 638°C (Sooksaen *et al.*, 2010). The DTA curves show that no clear endothermic peak was observed related to glass transition. However, an exothermic peak associated with the crystallization temperature of glass to form glass-ceramic was in the range of about 650°C to 850°C. Consequently, heat treatment was performed in this study at 700°C, 750°C, 800°C and 850°C to form the glass-ceramics, assuming phase evolution was completed at each isothermal hold of 1.0 to 12.0 hours.

Table 13.1: The Chemical Compositions of Raw Silica Sand

Oxide	SiO_2	Al_2O_3	K_2O	CaO	Na_2O	Fe_2O_3	MgO	TiO_2	LOI
Per cent	99.30	0.10	0.01	0.11	0.01	0.02	0.01	0.08	0.35

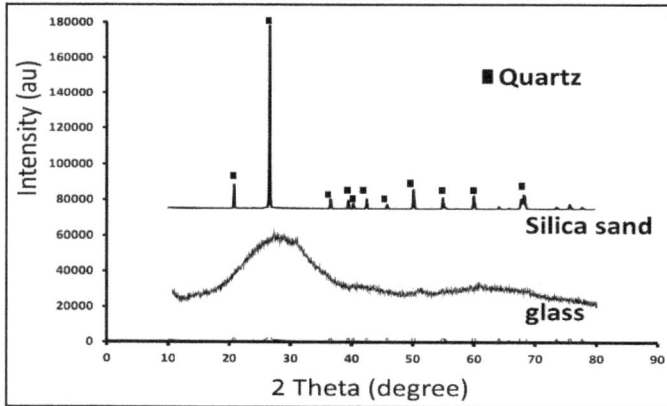

Figure 13.1: The Powder XRD Patterns of the Raw Silica Sand and Base Glass Sample

The glass-ceramics appearance after sintering at 700°C was slightly translucent and white. Increase in the sintering temperature up to 800°C caused the appearance of samples to become more white and opaque. The XRD patterns of glasses which were heat-treated at 700°C, 750°C, 800°C and 850°C for 1.0 hour duration are shown in Figure 13.3. The results indicated the presence of cubic leucite (ICDD: 00-038-1423) and sanidine (ICDD: 00-019-1227). It is evident that the amount of leucite and sanidine phase in each heat-treated glass composition increased with increasing

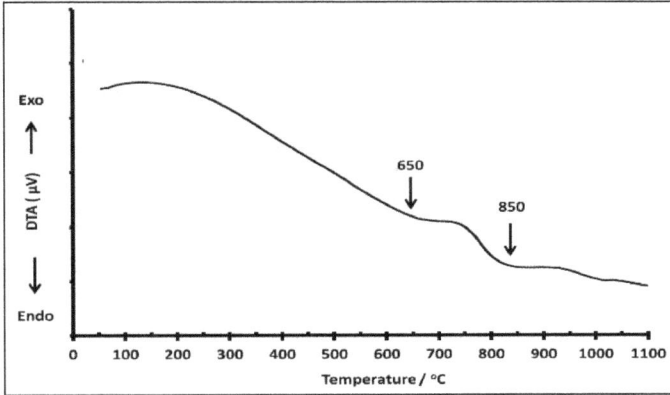

Figure 13.2: The DTA Curves for Base Glass Sample

heat treatment temperature. This is shown by increase in relative peak heights responsible for the both phases.

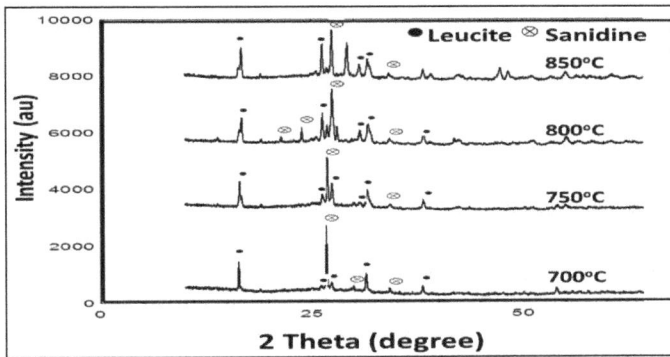

Figure 13.3: The XRD Pattern of Glass-Ceramics after Sintered at 700°C, 750°C, 800°C and 850°C for 1.0 hour Duration Time

The effect of sintering duration on the crystallization is shown in Figure 13.4. As the sintering duration was increased to 3.0, 6.0, 9.0 and 12.0 hours, all peaks associated with cubic leucite and sanidine became stronger. However, for sintering at 700°C and 750°C with soaking time of 6 hours, microcline (ICDD: 00-019-0932) started to form as a minor phase. While for 800°C and 850°C microcline started to crystallize within 3.0 hours soaking time. The crystallization of microcline was previously identified in leucite reinforced glass-ceramics in SiO_2-Al_2O_3-K_2O glass system (Cattell *et al.*, 2005).

Figure 13.5 shows the FESEM micrographs of glass-ceramics sample surface after sintering at 850°C for 1.0 hour and 9.0 hours. The formation of dendritic cubic leucite and prismatic sanidine phases is clearly seen. A previous study on crystallization mechanisms in glass-ceramics showed that the early stages of bulk leucite growth have been due to dendrites growing in preferred crystallographic

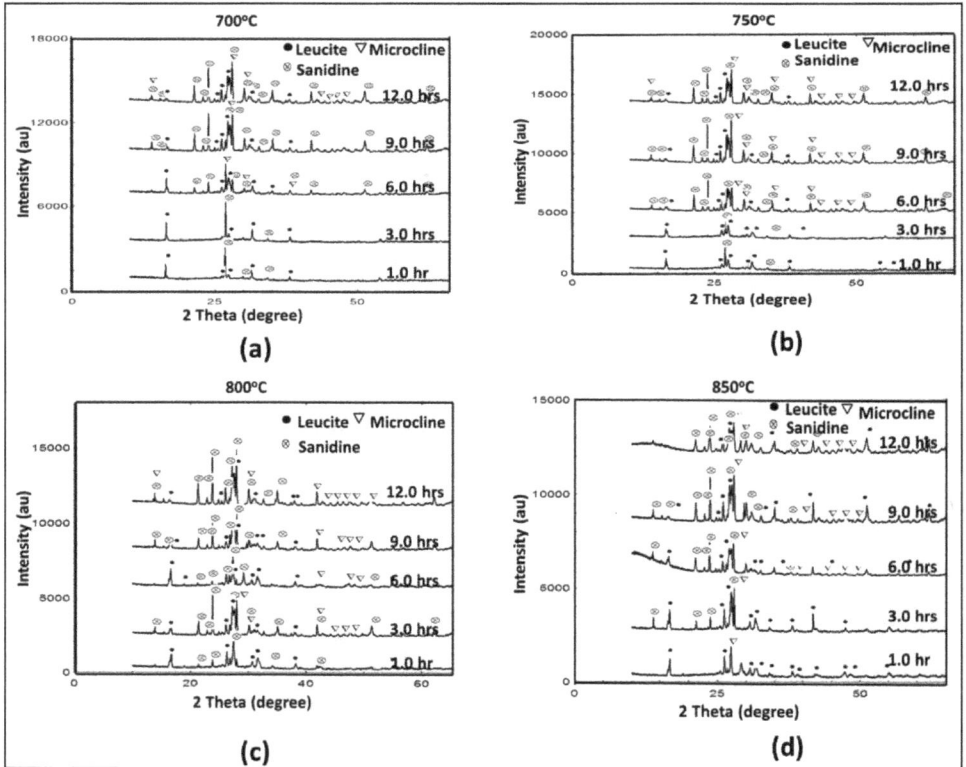

Figure 13.4: The XRD Patterns of Glass-Ceramics Aample after Sintering for (a) 700°C (b) 750°C (c) 800°C and (d) 850°C at Soaking Time of 1.0, 3.0, 6.0, 9.0 and 12.0 Hours

directions (Holand and Beall, 2002). The micrographs also show there are no obvious micro cracks and the surface is having a glassy texture.

Figure 13.6 shows the graph of the flexural strength results of the glass-ceramics samples after sintering at 700°C, 750°C, 800°C and 850°C and soaking at 1.0, 3.0, 6.0, 9.0 and 12.0 hours. The flexure strength of sintered leucite glass-ceramics increased sintering temperature and soaking time increased probably due to the high volume of crystalline phases away and the existence of prismatic sanidine. For example, the flexural strength of sample sintered at 700°C for 1.0 h was 80.0MPa and with increase temperature to 850°C the strength value increased to 120MPa. The higher flexural strength, 175MPa was achieved for sample sintered at 850°C for 9.0 hours. However, the flexural strength decreased after soaking for 9.0 hours at all temperatures. This is probably due to the glass-ceramics starting to melt and part of crystallite beginning to dissolve in the residual glass phase. The flexural strength values of sintered leucite glass-ceramics samples were comparable with commercial product of 112-140 MPa (El-Meliegy and Noort, 2011; Holand and Beall, 2002).

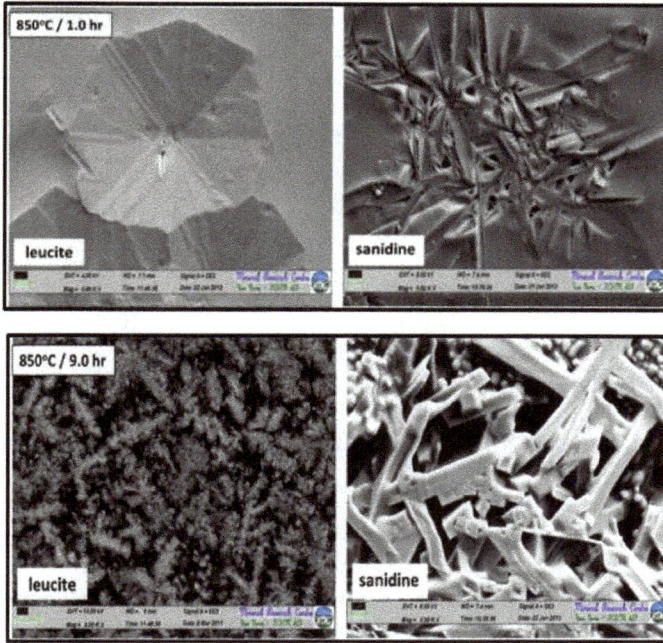

Figure 13.5: The FESEM Micrographs of Glass-Ceramics Sample Surface after Heat Treatment at 850°C for 1.0 and 9.0 Hours Show the Dendritic Cubic Leucite and Prismatic Sanidine Phases

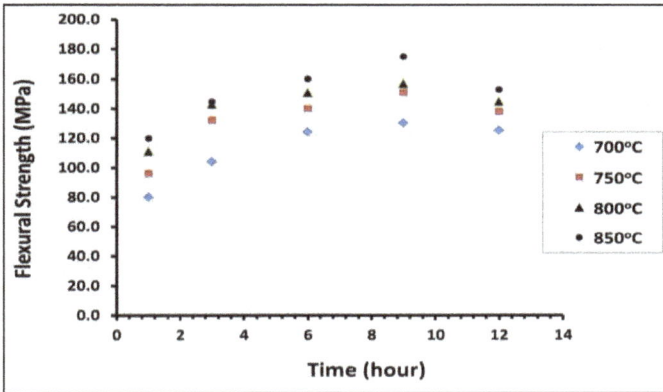

Figure 13.6: Flexural Strength as Function of Time for the Glass-Ceramics Samples after Heat Treatment at 700°C, 750°C, 800°C and 850°C

Glass-ceramics for restorative dental applications must fulfil the standard tests for biomaterial use, such as compatibility with the oral environment. Bioactivity on the surface of the dental restoration must not occur (El-Meliegy and Noort, 2011). XRD patterns of sintered glass-ceramics before and after soaking in SBF for several days are shown in Figure 13.7. The patterns show that the absence of apatite layer on the surface of glass-ceramics sample after 20 days of immersion in SBF indicating that the material is inert bioactive in nature.

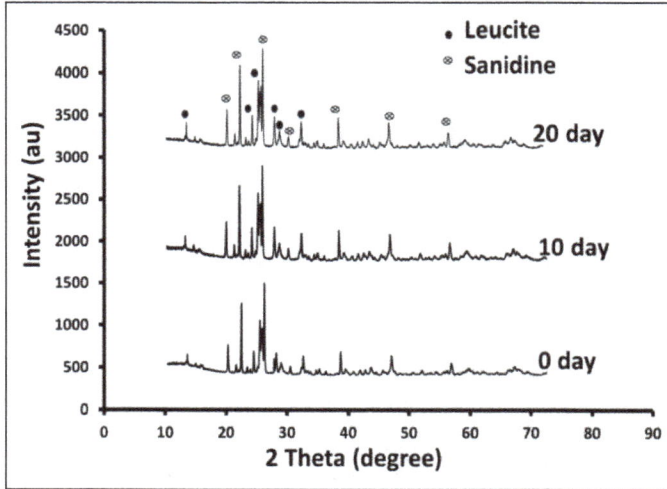

Figure 13.7: XRD Pattern of Sintered Glass-Ceramics Samples Before and After Soaking in SBF for 10 and 20 Days

The potential ability of the extracts of the sintered glass-ceramics material in inducing toxic effects on cell line at various concentrations was tested, the results of which are shown in Figure 13.8. MTT assay evaluated that the glass-ceramics sample did not inhibit the viability of L929 cells at all concentrations following 24 hours treatment. Both negative (complete growth medium) and positive control (zinc sulphate) performed as anticipated. If the material is not cytotoxic, the cells will remain attached and will proliferate in time. If the material is cytotoxic, the cells stop their proliferation showing cyto-pathologic features. In Figure 9, it shows that at all concentrations of the sintered glass-ceramics sample (including the negative control) the viability of the treated cell did not have so much difference.

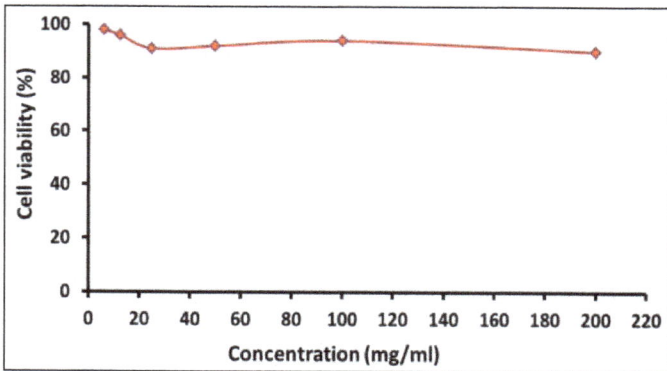

Figure 13.8: Viability of L929 Cells at Various Concentrations of the Test Glass-Ceramics Material

This proved that the glass-ceramics sample did not demonstrate α cytotoxic effect at all concentrations under the condition of study.

Conclusions

The highly pure natural Malaysian silica sand can be used as SiO_2 source for producing leucite glass-ceramics without any further chemical upgrading. Sintering of SiO_2-Al_2O_3-K_2O glass powder at 700°C to 850°C for duration of 1 to 12 hours contributed to the crystallization of cubic leucite and sanidine with minor phases of microcline. The leucite glass-ceramics has a flexural strength comparable with commercial product. The absence of apatite layer on the surface indicated that it is inert bioactive material. The leucite glass-ceramics is ranked non-cytotoxic in terms of in-vitro cellular response to human cell lines under the prevailing test conditions.

Acknowledgements

The authors would like to thank the Director of Mineral Research Centre, Department of Minerals and Geoscience Malaysia and fellow staff of the centre especially to the staff of Advanced Material Technology Section for their assistance in this research.

Refernces

1. Ahmad N., 2006. Production of high purity silica from Malaysian silica sand. Phd Thesis, University of Leeds, UK.

2. Cattel, M.J. Knowles, J.C. and Clarke, R.L. 2005. The crystallization of an aluminosilicate glass in the K_2O-Al_2O_3-SiO_2 system. J. Dental Materials, 21, pp. 811-822.

3. El-Meliegy, E.M. and Noort, R. 2011. Glasses and glass-ceramics for medical applications. Springer, New York, USA.

4. Holand, W. and Beall, G.H. 2002. Glass-ceramic technology. American Ceramic Society, Ohio, USA.

5. Holand,W. Frank, M. Rheinberger, V. 1995. Surface crystallization of leucite in glass. J. Non-Cryst. Solids. 180, pp. 292-307.

6. JMG, Malaysian Minerals Yearbook 2012. Ministry of Natural Resources and Environment Malaysia.

7. Malaysia Standard, MS701. 1981. Specification of Silica Sand for Glass Making. SIRIM, Malaysia.

8. Sooksaen, P. Boonmee, J. Witpathomwong, C. and Likhitlert, S. 2010. Effect of K_2O/SiO_2 ratio on crystallization of leucite in silicate based-glasses, J. Metals, Mater. and Minerals. 20(1), pp. 11-19.

Chapter 14

Fundamental Surface Properties of Carrollite (CoCu$_2$S$_4$) and its Flotation Behaviour

L.K. Witika[1], B. Musuku[1] and T. Hirajima[2]

[1]*Department of Metallurgy and Mineral Processing,*
School of Mines, University of Zambia, Box 32379, Lusaka, Zambia
E-mail: lwitika@unza.zm
[2]*Autokumpu Minerals Research Centre,*
Helsinki, Finland
[3]*Department of Earth Resources Engineering,*
Faculty of Engineering, Kyushu University, 744 Motooka, Nishiku,
Fukuoka 819-0395, Japan

ABSTRACT

Carrollite, known to be the main source of Cobalt, from the Zambian copper- cobalt bearing ores is a strategic mineral which when processed into a pure Cobalt metal can avail Zambia good foreign exchange revenue

In an effort to understand the surface properties of Carrollite and it's floatation behaviour, Electrokinetic measurements, Micro-Floatation experiment, XPS experiments and Contact Angles measurements where conducted on freshly prepared samples of the mineral.

Carrollite is fairly a stable double sulphide mineral (CuS.Co$_2$S$_3$) and shows little surface modification. After treatment with oxygen saturated solutions for 1hour, the surface charge as reviewed by zeta potential measurements, remains negative throughout the pH range 3 – 11. However, in oxygen saturated solutions its negativity is reduced as pH increases resulting in a charge reversal at about pH 7.5. Beyond pH 9 its negativity increases resulting into another charge reversal at about pH 10.5. A fresh Carrollite mineral sample has the following oxidation states for its elemental constituents; Cu (I), Co (III) and S (I and II).

The Carrollite mineral is reasonably stable at pH 3 in water, SIPX and oxygen saturated solution. Cu(OH)$_2$, CuS and CuX$_2$ are formed at high pH values (pH 9 and 11). The formation of hydrophobic species is optimum at pH 9. The maximum contact angles (33.5° and 49.8°) were achieved at pH 9 both for untreated and treated in SIPX polished sample respectively. This is confirmed in the high recoveries and achieved at pH 9 – 10.5. Collectorless flotation using Hallimond tube showed poor recoveries throughout pH range except for pH 3. Sodium Isopropyl Xanthate (SIPX) improved the recovery with maximum recovery of 95.21 per cent attained at pH 9.

Keywords: Sulphide minerals, Carrollite, Zeta potentials, Collectorless flotation, Oxidation states, Contact angles.

Introduction

Carrollite is a double sulphide mineral of Copper and Cobalt ($CuS.Co_2S_3$) that occurs along the Democratic Republic of Congo (DRC) – Zambian copperbelt region and is usually associated with other copper sulphide minerals such as Bornite (Cu_5FeS_4), Chalcocite (Cu_2S), chalcopyrite ($CuFeS_2$) and Cobaltiferous pyrite ((Co, Fe) S_2). However, Linneaite (Co_3S_4) is also known to be found in appreciable amounts in the DRC ore (Ngongo, 1975). Oxide minerals are mostly dominant in Ores from the open pit mines. The Zambian ores from the copperbelt region have a substantial amount of oxide minerals such as : Malachite ($Cu_2(OH)_2Co_3$), Pseudo-Malachite($Cu_5(PO_4)_2(OH)_4$), Azurite ($Cu_2(Co_3)_2(OH)_2$) and Chrysocolla ($CuSiO_3nH_2O$). The main sources of cobalt in Zambia are Carrollite and Cobaltiferous pyrite; with the former contributing about 95 per cent relative abundance. Konkola Copper Mines (KCM) has been facing challenges in improving the recoveries of Cobalt metal in their concentrators, which prompted them to propose this study in a bid to further understand the fundamental surface properties of Carrollite in relationship to it's flotation response. Literature on Carrollite is still in its infancy and more efforts need to be pursued to fully understand the various properties which can be exploited to improve it's recovery from other minerals. There are, however a lot of works on other sulphide minerals that can be used to understand their electrochemical and electrokinetic behaviours, which are cardinal for their flotation response (Salamy and Noxon, 1953).

It is generally recognized that most sulphide minerals are readily floated though the reasons may differ considerably. Some sulphide minerals can be floated without collectors whereas others require a small quantity of a collector (starvation amounts). Whether a sulphide mineral surface is naturally hydrophobic or not, is still open to discussion. However, it is generally accepted that apart from minerals such as Molybedenite, Orpiment, Realgar and Stibnite, pure and clean sulphide minerals are hydrophilic (Chander,1988; Miller, 1988). Flotation of various minerals may be classified in four categories depending on the type of minerals involved. However, in the case of sulphide minerals, three forms may be identified.

☆ Collectorless floatation in which certain hydrophobic surface species on mineral lead to flotation in the absence of collectors

☆ Selective floatation in which micro-bubble maybe assumed to attach on specific mineral surfaces in the presence of both collector and frother.

☆ Differential/segregation flotation in which a bulk concentrate is first obtained in the roughing circuit and then specific minerals are differentially/preferentially floated in a subsequent stage.

Molybdenite is one of the few sulphide minerals which is naturally floatable. The basis for native floatability rests on the assumption that sulphide lattice ions are expected to be weakly hydrated and do not interact strongly with water molecules (Chen and Morris, 1972). Sulphide minerals react with oxygen and water. The charge transfer mechanism between sulphur and oxygen leads to various oxidation states of sulphur. Extremely low levels of oxygen in the order of 10-20 atm can establish an oxidizing potential (E_h). One of the significant findings of electrochemical research on sulphide minerals is the recognition that under certain conditions sulphides will be oxidized to elemental sulphur. Under certain conditions the sulphur reacts with sulphides to form poly-sulphide species or metal deficient sulphides as shown in equation 1.

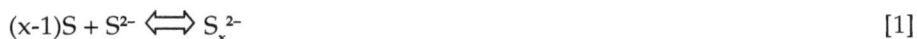

$$(x-1)S + S^{2-} \Longleftrightarrow S_x^{2-} \tag{1}$$

Polysulphides and metal deficient sulphides are usually formed at pH > 8 while elemental sulphur is dominant at pH < 6 (Chander, 1984). When these species are formed they render the mineral surface hydrophobic and hence a basis for collectorless flotation. This, however, should be distinguished from inherent floatability exhibited by a few sulphide minerals such as molybdenite, which is due to the composition and structure which makes them inherently hydrophobic (Gaudin *et al.*, 1957).

Oxidation of sulphide mineral surfaces is an important phenomenon, especially in the flotation process where it affects the attachment of collectors and may reduce the recovery and grade of valuable minerals. The occurrence of various reactions and their kinetics are strongly influenced by the potential difference across the mineral-solution interface (Chen and Morris, 1972). Thus the floatability of sulphide minerals excluding those which are believed to be naturally hydrophobic is greatly influenced by the degree of oxidation of the mineral surface. The presence of oxygen plays a pivotal role in ensuring that surfaces are adequately oxidized to aid floatation of the valuable minerals (Guy and Trahar, 1984, 1985).

Several kinds of electrochemical reactions involving sulphide minerals, dissolved oxygen and floatation reagents may occur depending on various thermodynamic conditions prevailing (Heyes and Trahar, 1986). The most common ones are:

Cathodic reduction of oxygen which may proceed through an intermediate reaction involving H_2O_2

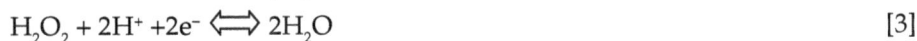

$$O2 + 2H^+ + 2e^- \Longleftrightarrow H_2O_2 \tag{2}$$

$$H_2O_2 + 2H^+ + 2e^- \Longleftrightarrow 2H_2O \tag{3}$$

To give the overall reaction

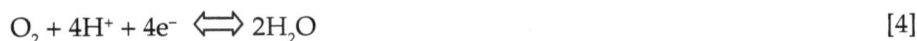

$$O_2 + 4H^+ + 4e^- \Longleftrightarrow 2H_2O \tag{4}$$

Anodic oxidation of collector through an adsorption phase

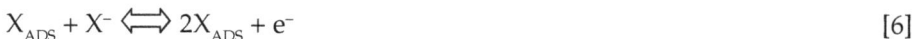

$$X^- \Longleftrightarrow X_{ADS} + e^- \tag{5}$$

$$X_{ADS} + X^- \Longleftrightarrow 2X_{ADS} + e^- \tag{6}$$

To give an overall reaction

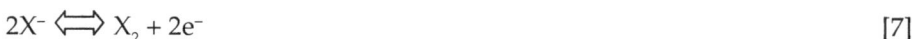

$$2X^- \Longleftrightarrow X_2 + 2e^- \tag{7}$$

Or

$$MS + 2X^- \Longleftrightarrow MX_2 + S + 2e^- \tag{8}$$

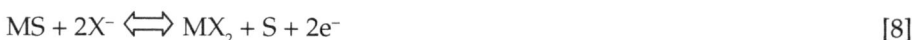

Sulphide mineral oxidation and formation of Polysulphides:

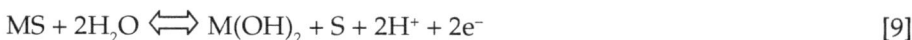

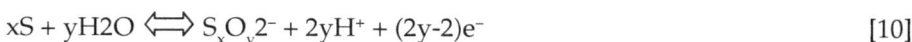

$$MS + 2H_2O \Longleftrightarrow M(OH)_2 + S + 2H^+ + 2e^- \tag{9}$$

$$xS + yH2O \Longleftrightarrow S_xO_y^{2-} + 2yH^+ + (2y-2)e^- \tag{10}$$

Equation 4 is the most important cathodic reaction and suggests that the presence of oxygen in sulphide systems is a pre-requisite for floatation to occur as was earlier postulated by other researchers (Guy and Trahar, 1984).

It is generally postulated that, the initial oxidation of simple sulphide minerals in acidic solutions may correspond to a reaction of the type (Chen and Morris, 1972) :

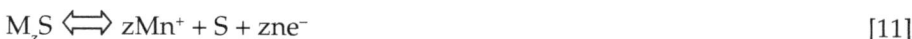

$$M_zS \Longleftrightarrow zMn^+ + S + zne^- \tag{11}$$

With its equivalent in neutral or alkaline solution as

$$M_zS + znH_2O \Longleftrightarrow zM(OH)n + S + znH^+ + zne^- \tag{12}$$

Oxidation leading to sulphur like products is usually considered to be critical in rendering some sulphide minerals such as chalcopyrite (Gaudin *et al.*, 1957) and galena (Guy and Trahar, 1984, 1985) naturally floatable (self induced floatable). However, oxidation leading to further oxidation products deteriorates this self induced floatability property. It implies that the nature of oxidation product is very critical more especially if they are soluble (Guy and Trahar, 1984).

Materials and Experimental Methods

Mineral Samples

Hand-picked natural almost pure samples of Carrollite were obtained from the Nchanga and Konkola ores of KCM. Elemental composition was determined using X-ray Fluorescence (XRF) whose results are shown in Table 14.1. Anal grade Sodium Isopropyl Xanthate (SIPX) was used as collectors, Potassium hydroxide (KOH) and Nitric acid (HNO$_3$) to regulate the pH and FZK 245 as frother. Pure water was used in the measurements of advancing contact angle. Electrokinetics, micro-flotation, XPS and contact angles were conducted at 25±10C. All experiments were done at Kyushu University, Department of Earth Resource Engineering, Laboratory of Mineral Processing and Recycling, Faculty of Engineering, Japan.

Table 14.1: Elemental Analysis of Carrollite

Mineral	XRF Results (Metal per cent)							
	Cu	Co	S	Si	Al	K	Ni	Fe
Carrollite	15.6	47.5	35.2	0.18	0.05	0.02	0.22	1.26

Carrollite samples were crushed with a porcelain mortar and sieved in a glove plastic to obtain -105μm to +78μm of size fraction for flotation and adsorption experiments and -38μm fractions for electrokinetic experiments.

Electrokinetic Measurement

-38μm Carrollite samples were suspended separately in 80ml of 10^{-3}M Potassium Chloride (KCl) at specific pH values and conditioned for 10, 30 and 60 minutes. After conditioning at the required pH values, the zeta potentials were measured using the electrophoretic machine equipped with a video camera. Pellets of SIPX were used as a collector with a mineral suspension of 6.25gpl. A total reading of 50 electrophoretic mobilities were recorded at first stationery level (0.15mm from the front rectangular cell wall) and converted to zeta potential using the ZEECOM ZC-2000system (Kyowa Interface Science Co., Ltd).

Micro-Floatation Experiments

Micro-floatation tests were carried out in a Hallimond tube at different pH values. A sample of +75μm to -106μm fraction was prepared in a 10^{-3} M KCl solution at different pH (3-11). The solution was conditioned for 10 minutes with and without SIPX before floatation at a constant stirrer rotational speed. During flotation nitrogen gas was used for aeration and the float was collected after 1minute of floatation. The tailings and float were dried and weighed. The recoveries were calculated based on the percentage weight of material floated of total weight of the sample (tailings + float).

$$R(\%) = \frac{C * 100}{(C + T)}$$

[13]

where,

☆ R is recovery (per cent)

☆ C is weight of floated material (g)

☆ T is weight of Tailings (g)

XPS Experiments

0.5g of Carrollite sample of -38μm particle size was suspended in 10^{-3}M KCl solution and conditioned for 1hour in 100ml Erlenmeyer flask in three different environments; without any additive, with oxygen being bubbled using a glass frit for producing small oxygen bubble with magnetic stirring and in the presence of SIPX at pH 3, 9 and 11.

After 1hour the mineral particles were separated by membrane filtration. Dissolved ions of Cu, Co and S concentrations were determined by Induced Coupled Plasma Atomic Emission Spectrometry (ICP-AES, Seiko Vista, Japan). Solid residues were dried overnight for analysis with X-ray photoelectron spectroscopy (PHI 5800 ESCA). The collected data were analysed with Casa XPS software (Ver. 2.3.12).

Contact Angles Measurements

Almost pure Carrollite specimens were mounted in epoxy to make briquettes which were then polished using polishing disc covered with different types of polishing clothes whose roughness ranged from 3000 to 1µm and using diamond paste as lubricant. The polished sections were then taken for equilibrium advancing contact angle measurements after pre-treatment for 10 minutes in acidic and alkaline media (pH 3- 11) using Drop Master Machine.

Results and Discussions

Zeta Potential Measurements of Carrollite in Aqueous Solution

Figure 1 shows that the zeta potentials of Carrollite in ordinary water were negative throughout the pH range. Conditioning for 10 and 30 minutes did not give significant difference except, for conditioning time of 60 minutes which showed slightly higher negativity. From this, it may be implied that the reaction governing this behaviour might have been the same and was function of time. From the general trend of the change in the zeta potentials with pH observed for elemental sulphur or non-oxidised sulphide minerals (Witika, 1988; Fornasiero *et al.*, 1999) one may logically conclude that, the sulphide mineral surface may be covered with sulphur like species which can become charged by reacting with protons in a similar manner to metal oxides. In the acidic media, the copper ions from the Carrollite lattice

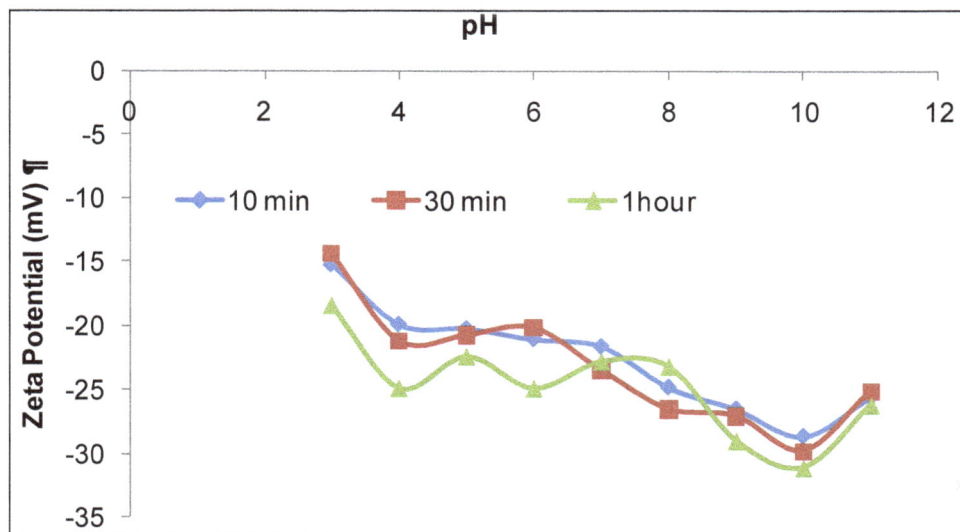

Figure 14.1: Zeta Potentials of Carrollite in Water as a Function of pH

structure went into solution. This behaviour is believed to be due to the presence of trace amounts of oxygen in water (Woods, 1984). The oxygen accepts electron from the copper ions thereby undergoing a cathodic reduction reaction as shown in equation 14.

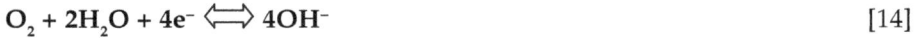

$$O_2 + 2H_2O + 4e^- \Longleftrightarrow 4OH^- \qquad [14]$$

The cathodic reaction causes the sulphide mineral to have a deficiency in metal ions and hence electrically unbalanced. As the metal ions go in solution, the overall surface charge is dominated by a negative charge which gives the mineral surface its negative charge. The oxy-hydroxide formed from cathodic reaction of oxygen coats the metal-deficient sulphide rich mineral surface at higher pH values hence giving it a more negative charge. As pH increases from acidic to alkalinity, the copper ions react with the oxy-hydroxide to form copper hydroxide (Lombe and Witika, 1989). As the pH increased beyond 8, the negativity increased up to pH 11. This was a result of dissolution of cobalt ions into solution and cathodic reaction of oxygen hence formation of oxy hydroxides which are negatively charged. The reactions may be represented by the following equations:

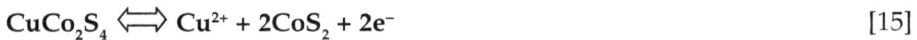

$$CuCo_2S_4 \Longleftrightarrow Cu^{2+} + 2CoS_2 + 2e^- \qquad [15]$$

CoS_2 is very unstable, therefore, it dissociates through the following step (Plaskin and Bessonov, 1957)

Dissolution of cobaltous ions to form a cobalt deficient sulphur rich surface rather than elemental sulphur

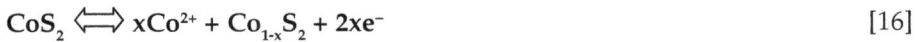

$$CoS_2 \Longleftrightarrow xCo^{2+} + Co_{1-x}S_2 + 2xe^- \qquad [16]$$

The oxidation of cobaltous ions to cobaltic ion at higher oxidation potential and pH

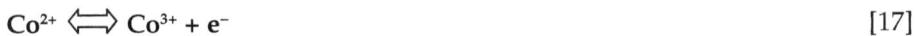

$$Co^{2+} \Longleftrightarrow Co^{3+} + e^- \qquad [17]$$

The hydrolysis of cobaltous hydroxide species whose concentration are largely dependent on solution pH;

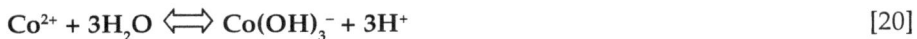

$$Co^{2+} + H_2O \Longleftrightarrow Co(OH)^+ + H^+ \qquad [18]$$

$$Co^{2+} + 2H_2O \Longleftrightarrow Co(OH)_{2(aq)} + 2H^+ \qquad [19]$$

$$Co^{2+} + 3H_2O \Longleftrightarrow Co(OH)_3^- + 3H^+ \qquad [20]$$

Finally the formation

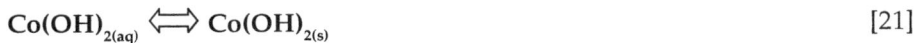

$$Co(OH)_{2(aq)} \Longleftrightarrow Co(OH)_{2(s)} \qquad [21]$$

Figure 14.2, shows that, the zeta potentials were more negative between pH 3 and somewhere just below pH 6 as a result of increased surface oxidation due to the presence of excess oxygen in the solution. This enhanced the cathodic reactions of oxygen, leading to higher amount of oxy-hydroxide anionic species forming, which coated on the surface of the mineral (Chen and Morris, 1972; Plaskin and Bessonov, 1957). This demonstrates the role of oxygen in determining zeta potentials of minerals. At higher pH, they were two charge reversals at pH 7.5 and 10.5. The

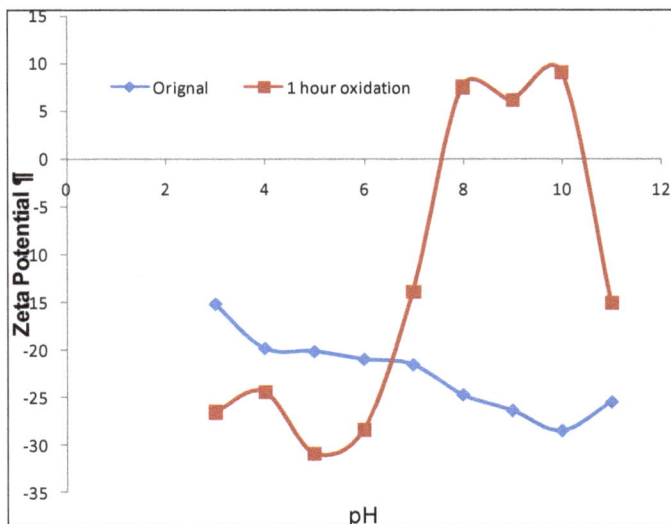

Figure 14.2: Zeta Potentials of Carrollite in Oxygen Saturated Solutions as a Function of pH

initial charge reversal can be attributed to the precipitation of copper hydroxide cationic like species which precipitated on the mineral surface. It is well documented that copper hydroxide cationic species precipitates at pH higher than 5 (Woods, 1984). The dissolution of cobalt ions into solution, leaves behind a positively charged metal deficient sulphide rich surface as some of the metal ions which are positively charged cover the mineral surface. As the pH increased, the cobalt ions formed cobalt hydroxides (eq. 17-21) and consequently reduced the positive zeta potential, causing a second charge reversal at pH 10.5.

Figure 14.3 shows the dissolution of carrollite as a function of pH. The ionic species found in solution at lower pH are mostly copper ions whereas at higher pH

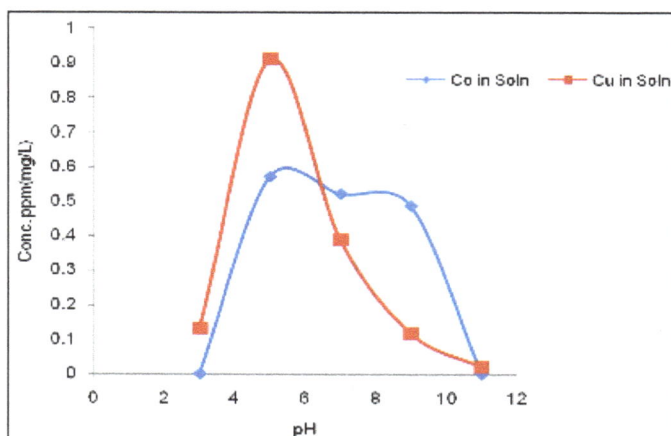

Figure 14.3: Dissolution of Carrollite in Aqueous Solution as a Function of pH

cobalt ions are present. This implies that most of cobalt ions can only go in solution around pH 6 and its hydroxides are formed at higher pH values. It can be deduced that copper ions play a pivotal role in determining zeta potential in acidic media whereas cobalt does so in alkaline media (Lombe and Witika, 1989).

Zeta Potential of Carrollite in SIPX Solution

Figure 14.4 shows that when SIPX goes in solutions, depending on the pH, xanthate ions are formed (Chander, 1988). These anions are believed to coat the mineral surface throughout pH range hence increasing the negative charge on the mineral surface.

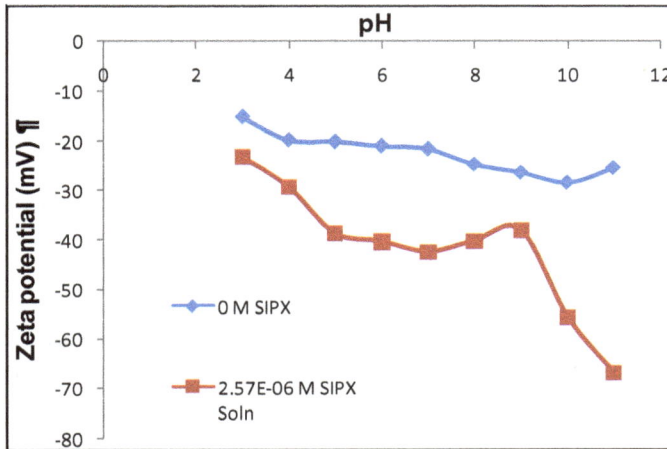

**Figure 14.4: Zeta Potentials of Carrollite in SIPX Saturated
Solutions as a Function of pH**

X-ray Photoelectron Spectroscopic studies.

Fresh Samples

Figure 14.5, shows that the binding energy (BE) for Cu $2p_{3/2}$ peak was observed to be 932.3 eV which is characteristic of Cu (I) which is within the range as reported elsewhere (Woods, 1984). Other researchers (Witika, 1996, Alan *et al.*, 2009) determined the Cu $2p_{3/2}$ binding energy for Carrollite to be 932.6 eV and 932.5 eV respectively for Cu (I). The 778 eV BE found in this work is characteristic of Co (III). The results from the Table 14.2, below for Co binding energy are found to be characteristic of Co (III). The peak separation between $2p_{3/2}$ and $2p_{1/2}$ was found to be 15.0 eV.

Treatment with Water SIPX and Oxygen

Figure 14.6 shows, the effect of water, SIPX and Oxygen at different pH values on the surface of the mineral. Carrollite was observed to be fairly stable at pH 3 in water, SIPX and Oxygen. This agrees very well with the amount of metal ions in solution. At pH 9, there was an increase in intensity at BE 933.8 eV, with the peak

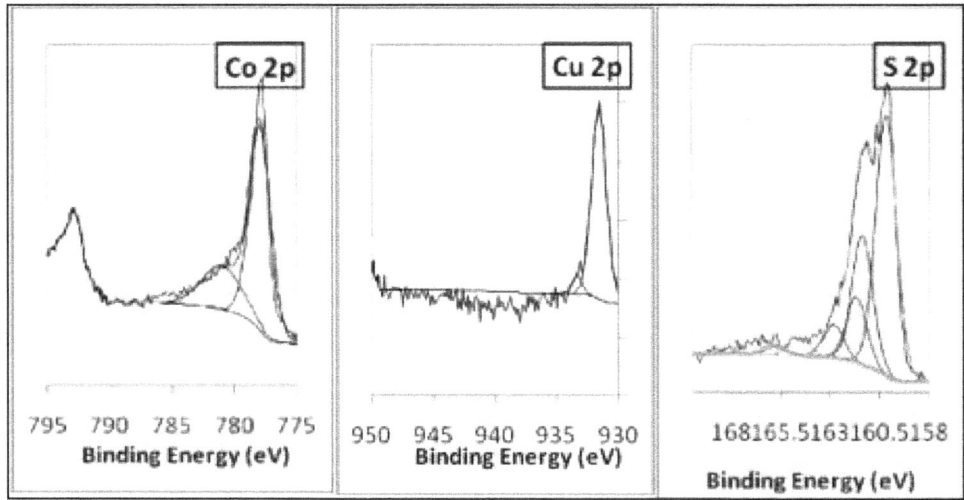

Figure 14.5: Cu 2p, Co 2p and S 2p Spectra for Untreated Carrollite Sample

Table 14.2: Co $2p_{3/2}$ and S $2p_{3/2}$ BE (eV) for Carrollite from different Works

Material	Source	$Co2p_{3/2}$ BE (eV)	$2p_{3/2}$–$2p_{1/2}$	S $2p_{3/2}$ BE (eV)	Reference
$CuCo_2S_4$	Baluba	~779	15	161.7	Witika, 1996
$CuCo_2S_4$	Katanga	779	15	161.2	Alan *et al.*, 2009
$CuCo_2S_4$	Nchanga	778	15	161.4	Musuku, 2012

The 161.4 eV binding energy for S $2p_{3/2}$, as observed in this work can be assigned to S (II) (Chen and Morris, 1982). The second peak at BE 161.7 eV can be assigned to S (I). Therefore, the oxidation state of Carrollite was found to be in the form $Cu^I Co^{III} S^{-II} S^{-I}$.

being more pronounced in oxygen environment. This was assigned to Cu (II) (Chen and Morris, 1982). The Cu is believed to be oxidized from Cu (I) to Cu (II).

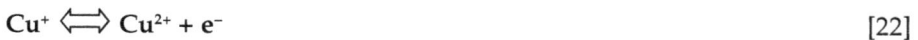

$$Cu^+ \rightleftharpoons Cu^{2+} + e^- \qquad [22]$$

Therefore, the most probable products to be formed at this pH would be $CuO/Cu(OH)_2/CuS$. Work done earlier (Witika, 1988) has concluded using cyclic voltammetry experiments that CuS is formed on the surface of Carrollite mineral in alkaline media;

$$CuS.Co_2S_3 + 4H_2O \rightleftharpoons CuS + 2Co(OH)_2 + 3S + 4H^+ + 4e^- \qquad [23]$$

CuO and $Cu(OH)_2$ was ruled out due to the fact that beyond pH 7 the two compounds are unstable and this was confirmed by XRD experiments and E_h – pH diagram (Witika, 1988). At pH 11 the peak for 933.8 eV, characteristic of Cu (II) was more pronounced.

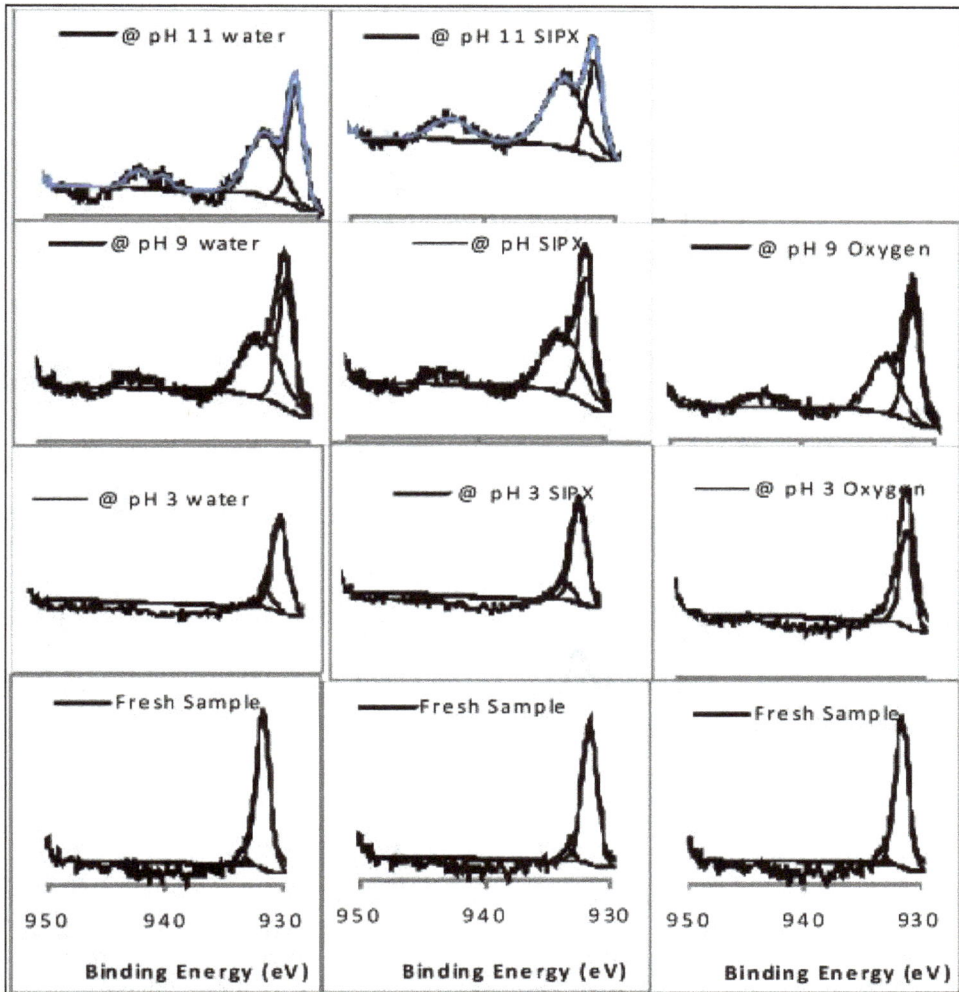

Figure 14.6: Cu 2p Spectra for Carrollite in Water, SIPX and Oxygen

Figure 14.7 shows the Co 2p$_{3/2}$ spectra before and after treatment with water, SIPX and in oxygen at pH 3, 9 and 11. The Co spectra do not seem to be affected at pH 3. This is in conformity with the amount of Co ions that went in solution at pH 3 as shown in Figure 3. A slight change on mineral surface is observed at pH 11. The binding energy of 781.0 eV slightly increased compared to original sample. This BE is characteristic of Co(II). This implies that in alkaline media, Co (III) is reduced to Co (II). The expected product at this pH is Co(OH)$_2$.

Figure 14.8, shows that, the S 2p spectra is in confirm of the presence of metal hydroxide, metal-deficient sulphide, polysulphide and elemental sulphur on the cobalt mineral. BE 163.0 and 164.2 eV indicate the presence of metal-deficient

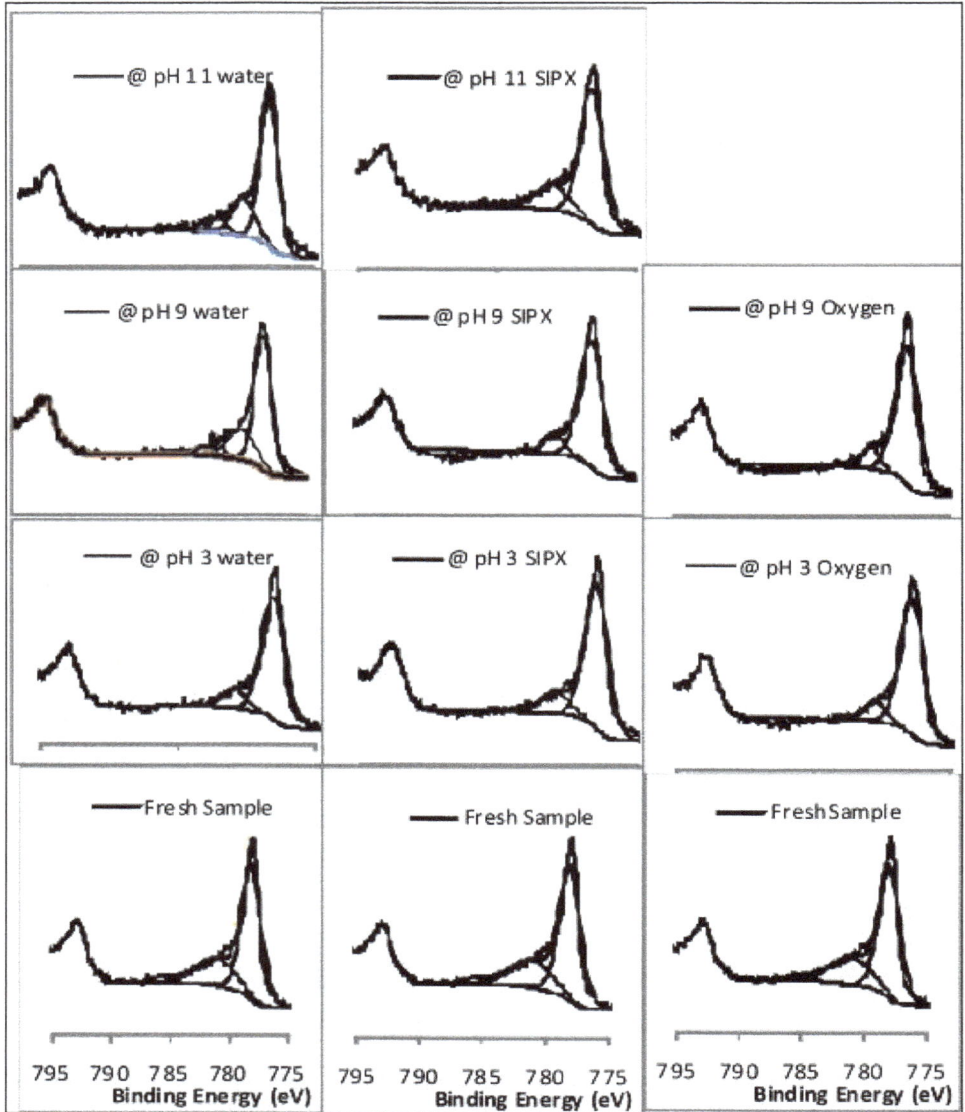

Figure 14.7: Co 2p Spectra for Carrollite in Water, SIPX and Oxygen

sulphide and poly sulphide respectively. These products are more pronounced at pH 9 and disappear at pH 11. The peak at BE 161.4 eV increases with pH which confirms the formation Cu (II) products.

Contact Angle Measurements

Contact angles are important parameters as they are commonly used as a measure of the hydrophobicity or wettability of a solid surfaces when in aqueous

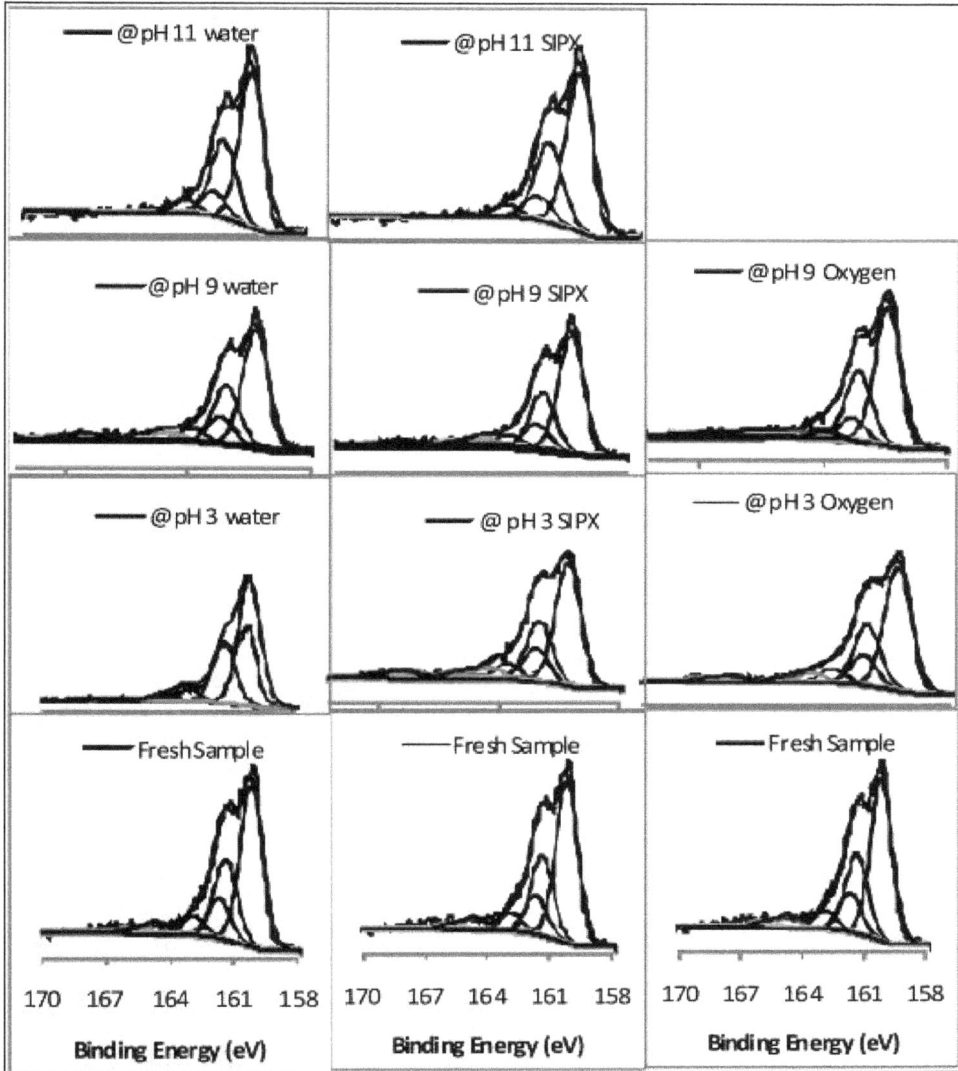

Figure 14.8: S 2p Spectra for Carrollite in Water, SIPX and Oxygen

solutions (Chau *et al.*, 2009). Figure 14.9, shows Advancing angles measurements on polished carrollite specimens as a function of pH.

Figure 14.9 shows that, for both untreated and treated with SIPX carrollite polished specimens, the measured contact angles follow a similar trend, increasing with the increase in pH. This implies that compounds responsible for hydrophobicity of a mineral surface were in appreciable amounts around pH range 8 to 9.2. XPS results, reported earlier, showed that at pH 9 there was evidence of metal deficient

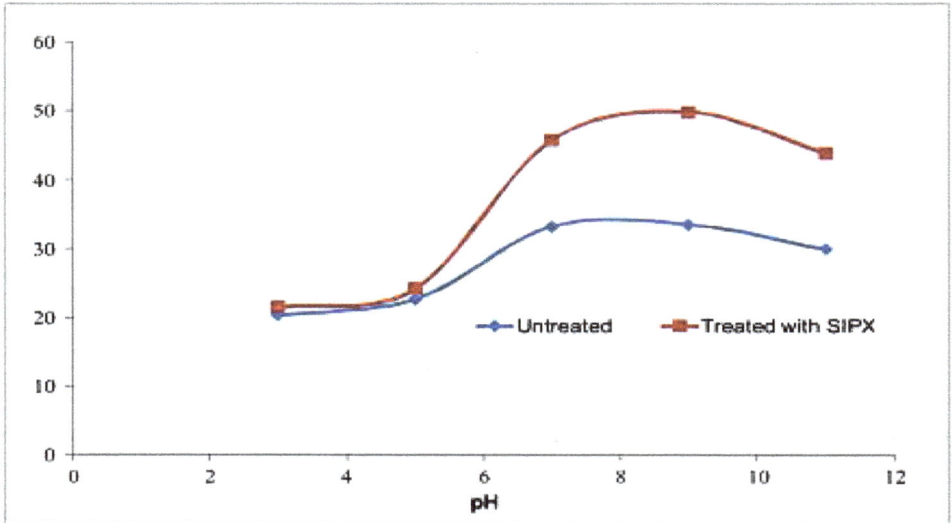

**Figure 14.9: Advancing Contact Angles Measurements on
Carrollite Polished Specimens as Function of pH**

sulphide, Cu_{1-x} S and poly sulphide on the surface of the treated mineral surface. Beyond pH 9 metal hydroxides were observed on the mineral surface. Therefore, from these results, it can be deduced that metal deficient like sulphide Cu_{1-x} S, poly sulphide and perhaps metal xanthates and dixanthogen, were responsible for rendering the mineral surface hydrophobic while the decrease in contact angle beyond pH 9 can be attributed to the presence of hydrophilic metal hydroxides like species on the mineral surface.

Micro-Flotation Experiments

Figure 14.10, shows that, the collectorless floatation of Carrollite was very poor as the pH increases above pH3. Addition of collector (SIPX) improved the recovery throughout all pH range perhaps due to combination of both the presence of metal deficient sulphides and elemental sulphur and formation of dixanthogen particularly in the pH range of 7 – 10. The recovery reduces beyond pH 10, which agrees with the presence of hydrophilic $Co(OH)_2$ on the mineral surface.

Figure 14.11, shows that the maximum recovery (95.21 per cent) was achieved at pH 9 with 32g/t SIPX dosage. Other dosages showed steady drop in recovery with increase in pH.

The low recoveries can be attributed to the amount of adsorption of Xanthate ions on the mineral surface. In the case of the lowest dosage (24g/t) the amount of Xanthate ions attaching on the mineral surface to induce hydrophobicity could have been not sufficient. However, the trend at pH 9 is similar to 32g/t, and it can be deduced from this that probably pH 9 is the optimum pH for the formation of species on the mineral surface responsible for hydrophobicity. Formation of dixanthogen, which is responsible for hydrophobicity, is favoured at about pH 9 as revealed by

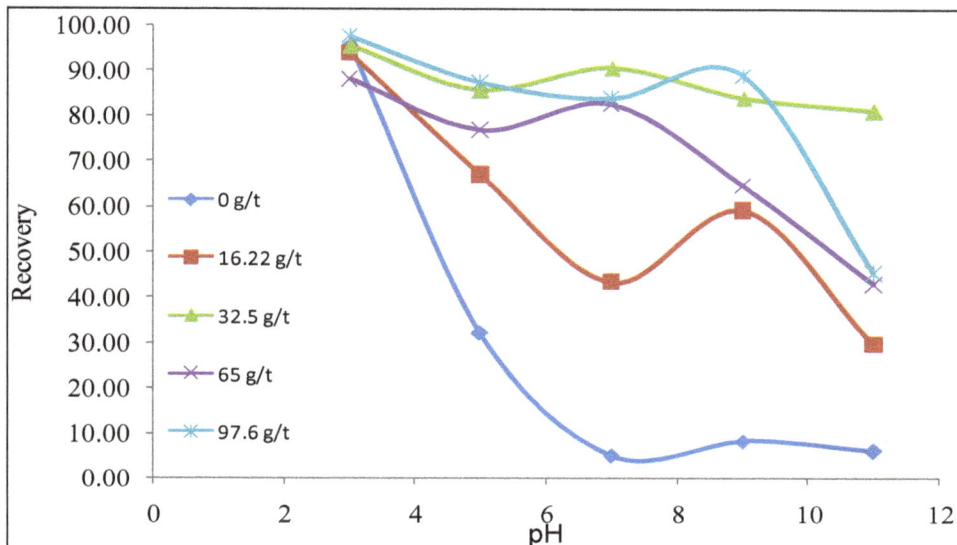

Figure 14.10: Hallimond Tube Flotation of Pure Carrollite Mineral Crystals

UV-Vis experiments on SIPX. However, at higher dosage (40 and 48g/t), it can be assumed that multilayer of the collector is adsorbed on the mineral surface. This situation re-orients the collector molecules, thus leading to the depression of the mineral.

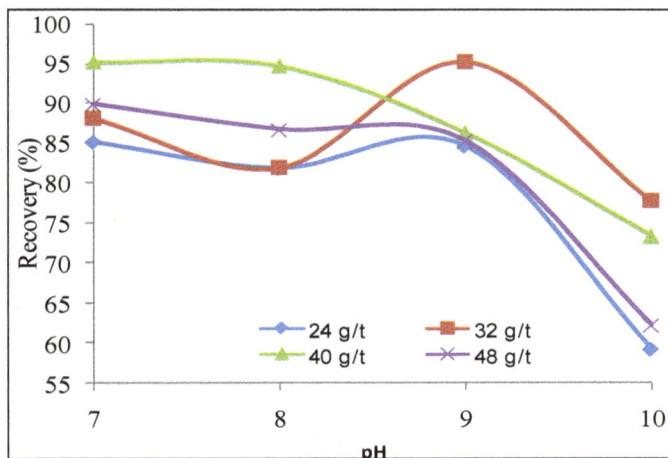

Figure 14.11: Recovery of Carrollite Minerals in the pH Range 7-10

Figure 14.12 shows that, the recovery was maximum at conditioning at pH 9. This may be attributed to the presence of xanthate ions which formed metal xanthate and dixanthogen at pH 9 as compared to other pH values, thereby inducing hydrophobicity and subsequently improving the recoveries giving 32g/t as the optimum dosage as earlier established.

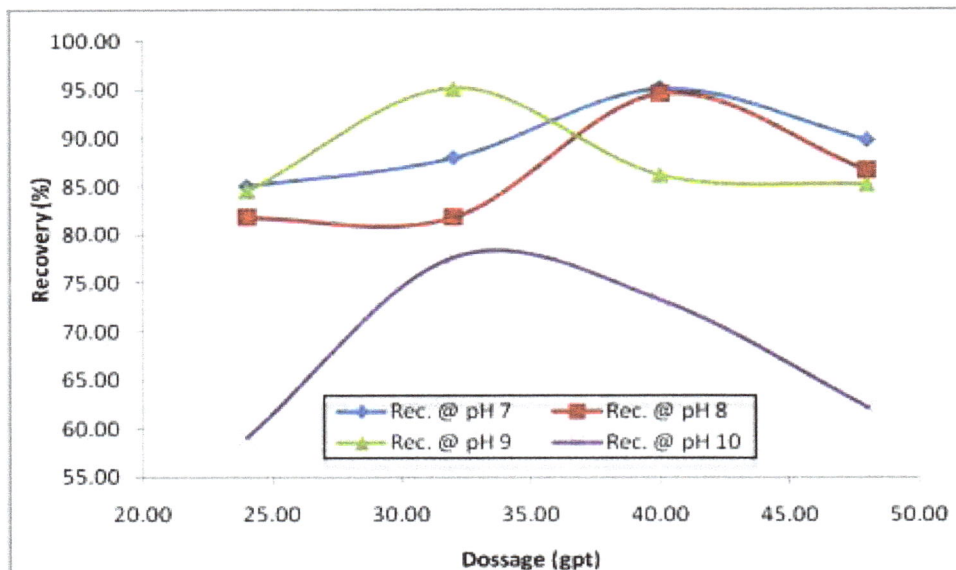

Figure 14.12: Recovery at different Conditioning pH as a Function of SIPX Concentration

Conclusions

Carrollite surface acquire a negative charge, when in aqueous solution in the pH range 3 – 11. The sulphide mineral loses electrons in the presence of oxygen in the water as the metal ions go in solution thereby forming oxy-hydroxides (OH^-) which covers the metal deficient sulphide (positively charged) surface. The presence of SIPX increases the negativity of the mineral surface due to the presence of Xanthate anions which preferentially adsorb on the mineral surface. Oxidation of the Carrollite causes charge reversal at about pH 7.5 and 10.5.

SIPX reacts with water to form xanthic acid at low pH. At pH 6 and above Xanthate ions are present in solution due to ionisation reaction. The presence of Xanthate ions in solution is a necessity to the formation of dixanthogen (X_2). Adsorption of Xanthate ions on mineral surface may lead to the formation of metal Xanthate/dixanthogen which may result in a monolayer of perhaps in the first 5minutes as determined by UV-Vis spectrophotometry measurement. A second layer may possibly be formed after 30 minutes conditioning time. However, a monolayer is adequate to induce the required hydrophobicity properties.

The oxidation states of copper, cobalt and sulphur in Carrollite fresh sample are Cu (I), Co (III) and S (I) S (II). Carrollite is fairly stable in acidic media, however, at pH 9 and 11 copper is oxidised to Cu (II) (CuS, $Cu(OH)_2$ and CuX_2) whereas cobalt is reduced to Co (II) ($Co(OH)_2$, CoS and CoX_2).

Contact angles at low pH were below **30°** while at pH 9, the maximum angles attained were **33.5°** and **49.8°** for carrollite minerals untreated and treated with SIPX respectively. This implies that favourable hydrophobic species may be formed at pH 9 and SIPX merely improved the hydrophobicity of the mineral.

Acknowledgements

The authors would like sincerely thank Konkola Copper Mines plc, for supporting BM to conduct some studies at Kyushu University in the Department of Earth Resources Engineering Faculty of Engineering in Japan.

References

1. Alan N.B., William M.S. and Harmer S.L., "Electronic environments in Carrollite, determined by soft X-ray photoelectron and absorption spectroscopy. Science Direct. *Geochinmica et Cosmoschimica Acta* 73, 4452 – 4467, 2009.

2. Chander S.," Electrochemistry of Sulphide Minerals Flotation". Minerals and Metallugical Processing, pp 104-114.1988.

3. Chau T.T., Bruchard W.J., Koh P.T.L. and Ngunyen A.V., "A review of Factors affecting Contact Angles and Implications for Flotation Practice." Advances in Colloid and Interface Science, 150, 106-115, 2009.

4. Chen K.Y. and Morris J.C., "Kinetics of Oxidation of Aqueous Sulphide by Oxygen". Environmental Sci. Technol, 6, pp 529-537, 1972.

5. Fornasiero D., Fullston D. and Ralston J., "Zeta potential study of oxidation of copper sulphide minerals". Colloids and surfaces, 146, 113-121, 1999.

6. Gaudin A.M., Miaw K.L. and Spedden H.R. "Native Floatability and Crystal Structure". "2nd International Congress of Surface Activity, Ed. J.H. Schulman, Butterworth, London, pp 201-219, 1957.

7. Guy, P. J. And Trahar, W. J., "The effect of oxidation and mineral interaction of sulphide flotation." Development in Mineral Processing, Volume 6, S. K. E. Forrsberg (Editor) Elsevier, Amsterdam, 1985

8. Guy P.J. and Trahar W.J., "The Influence of Grinding and Flotation Environment on the laboratory batch Floatation of galena". International Journal of Mineral Processing,12: 15-38, 1984.

9. Heyes G.W. and Trahar W.J., "Oxidation Reduction Effects in Flotation of Chalcopyrite and Cuprite". International Journal of Mineral Processing, 1986.

10. Heyes G.W. and Trahar W.J., "The Natural floatability of Chalcopyrite". International Journal of Mineral Processing, 4, 317-344, 1977.

11. Lombe, W. C. and Witika, L. K., " Electrochemical behaviour of carrollite in aqueous buffer solutions" Transaction of Institution of Mining and Metallurgy, London, Jan-April, C26-32., 1989.

12. Miller, J. D., " The significance of electrochemistry in the analysis of mineral processing phenomena" 7th Australian Electrochemistry Conference, Sydney, Australia, pp 14-19, February, 1988.

13. Musuku, B," Fundamentals of surface properties of Carrollite." MSc thesis, University of Zambia, 2012.

14. Ngongo, N., " Sur la Similitude entre les gisements uraniferes (type Shinkolobwe) et les gesiments cupriferes (type Kamoto) au Shaba, Zaire" Annales de la Societe Geoloque de Belgigue, Vol. 98, pp 449-463, 1975.

15. Plaskin I.N. and Bessonov S.V.,." Role of Gases in Floatation reactions". 2nd International Congress on Surface Activity, 3, 361-367, 1957.

16. Salamy, S. G. and Nixon, J. C." The application of electrochemical methods to floatation research." In : Recent developments in Mineral Dressing. Inst. Mini., and Metallurgy, London, pp 503-516, 1953

17. Sasaki K., Takatsugi K., Ishikura K and Hirajima T., "Spectroscopic study on oxidation dissolution of chalcopyrite, Enargite and Tennantite at different pH values" Hydrometallurgy 100, 144 – 151, 2010.

18. Witika, L. K., " X- ray photoelectron spectroscopy (XPS) analysis of carrollite". In: Proceedings of the 3rd International Conference on Mineral Processing and Environment, 392-402, P. Fecko (Ed.), Ostrava, Czeck Republic, 1996.

19. Witika, L. K. And Dobias, B., " Electrokinetics of Sulphide Minerals: Fundamental surface reactions on carrollite". Mineral Engineering, Vol. 6, No 8-10, pp 883-894, 1993

20. Witika, L. K. " Investigations of the Electrochemical, Electrokinetic and Adsorption of Carrollite and Chalcopyrite from Zambia in relationship to their flotation behaviour." DSc Thesis, University of Regensburg, Germany, 1996.

21. Lombe W. C. and Witika, L. K, " Electrochemical behaviour of Carrollite $CoCu_2SO_4$) in aqueous solutions. Transaction of the Institution of Mining and Metallurgy, London, Jan- April, 1989, C26-32.

22. Woods R., "Electrochemistry of sulphide flotation". In "Principles of Mineral Flotation". Ed.M.H. Jones, J.T. Woodcock. The Australasian Institute of Mining and Metallurgy, Victoria, Australia, 91–115, 1984.

Harare Resolutions on Minerals Processing and Beneficiation

WHILE EXPRESSING GRATITUDE to His Excellency the President of the Republic of Zimbabwe, Robert G. Mugabe, for presiding over the official opening ceremony of the 3rd International Workshop on 'Minerals Processing and Beneficiation' on the 11th September 2014, during which he appreciated the Centre for Science and Technology of the Non-aligned and Other Developing Countries (NAM S&T Centre) as a key thrust and a platform for advancing the developmental imperatives of the NAM and Other Developing Countries through cooperation in innovation, trade, beneficiation and value addition of their mineral endowment, and offered to host the proposed Non-aligned Movement Science and Technology Centre of Excellence for Mineral Processing and Beneficiation;

NOTING WITH CONCERN the array of initiatives from/by developed countries aimed at securing unfettered access to raw minerals from developing countries while discouraging the same developing countries' efforts for beneficiation and value addition;

NOTING WITH APPRECIATION the central role and guidance rendered by the Republic of Zimbabwe through the Ministries of Higher and Tertiary Education, Science and Technology Development; and Mines and Mining Development jointly with the NAM S&T Centre in organising and hosting the international workshop;

FURTHER NOTING the contribution of more than 110 participants from 16 NAM and Other Developing Countries; and the presentation of 22 high quality technical papers;

HAVING DELIBERATED on mineral resource endowment, policies, strategies, regulatory frameworks, research, innovation as well as technological interventions in support of mineral processing and beneficiation in developing countries;

RECOGNISING the challenges faced by NAM and Other Developing Countries in mineral development, processing and beneficiation, including issues of environmental sustainability, technology transfer, skills and infrastructure development;

NOTING that actual mineral wealth can be generated by establishment of designated minerals market places and their commodity exchange mechanisms as enablers of value addition, investment, and strengthening the role of developing countries in international trade of their minerals;

WE, THE PARTICIPANTS OF THE WORKSHOP, representing Afghanistan, Guyana, Indonesia, Iran, Malaysia, Mauritius, Myanmar, Nigeria, South Africa, Sri Lanka, Tanzania, Uganda, Vietnam, Zambia and Zimbabwe unanimously resolve to:

(i) Immediately undertake the process of establishing the Centre of Excellence for Mineral Processing and Beneficiation in Zimbabwe;

(ii) Establish a taskforce, as part of the above Centre, for preparing draft policy guidelines and legal frameworks for designated mineral market places and commodity exchanges within developing countries to be presented to African Union Council of Ministers responsible for minerals within the context of the African Mining Vision and AU agenda 2063 and subsequently to the AU Heads of State Summit; as well as the NAM Heads of States Summit;

(iii) Establish new research and development institutes and strengthen the existing ones for capacity building and mobilise adequate funding for mineral processing and beneficiation in NAM and Other Developing Countries;

(iv) Strengthen collaboration among NAM and Other Developing Countries in the setting up and implementation of sustainable and appropriate Minerals Development, Processing and Beneficiation Policies including regulatory frameworks;

(v) Promote value addition of mineral resources and resource-based industrialization through the adoption of sustainable and time bound projects and programmes;

(vi) Encourage public social private partnerships (PSPP) for the development and commercialisation of new and emerging technologies and ensuring the role of entrepreneurs and youth ventures;

(vii) Promote strategic Human Resource Development in mineral processing and beneficiation and encourage information exchange of scientists and technologists among NAM and Other Developing Countries;

(viii) Facilitate access to high-tech research infrastructure, and international collaboration among NAM and Other Developing Countries and promote localised uptake of innovative technologies;

(ix) Strongly recommend that Governments of Developing Countries refrain from engaging foreign Non-state actors including Non-Government Organisations in the governance of their mineral resources.

It was proposed by the delegate of Uganda to host the next international workshop on this theme sometime in 2016 jointly with the NAM S&T Centre, subject to the availability of funds and necessary government approvals. The participants of the workshop expressed gratitude to them for this kind gesture.

THUS, RESOLVED IN HARARE, REPUBLIC OF ZIMBABWE ON THIS DAY, 13th SEPTEMBER 2014.

www.ingramcontent.com/pod-product-compliance
Lightning Source LLC
Chambersburg PA
CBHW050515190326
41458CB00005B/1543